THATCH
Thatching in England 1940–1994

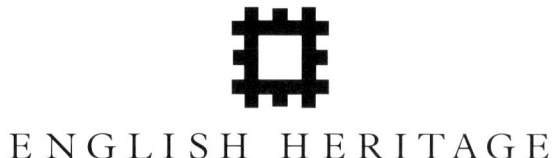

ENGLISH HERITAGE

ENGLISH HERITAGE RESEARCH TRANSACTIONS

RESEARCH AND CASE STUDIES IN ARCHITECTURAL CONSERVATION

THATCH

Thatching in England 1940–1994

Jo Cox & John Letts

Volume **6**

April 2000

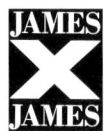

© 2000 English Heritage (text)
© 2000 illustrations and photographs: copyright of the authors
or sources cited in the captions

All rights reserved. No part of this publication may be reproduced,
stored in a retrieval system or transmitted in any form or by any
means, electronic, mechanical photocopying, recording or otherwise,
without the prior written permission of the copyright owner
and the publisher.

Published by James & James (Science Publishers) Ltd,
35-37 William Road,
London NW1 3ER, UK

A catalogue record for this book is available from the British Library
ISBN 1-873936-96-6
ISSN 1461 8613

Authors: Jo Cox and John Letts
Volume editor: Nicholas Molyneux
Series editor: David Mason
Consultant editor: Kate Macdonald

Typeset by James & James (Science Publishers) Ltd
Printed in the UK by The Alden Press Ltd, Osney Mead, Oxford

Disclaimer
Unless otherwise stated, the conservation treatments and repair
methodologies reported in this volume are not intended as specifications for remedial work. English Heritage, its agents and publisher
cannot be held responsible for any misuse or misapplication of
information contained in this publication.
The inclusion of the name of any company, group or individual, or of
any product or service in this publication should not be regarded as
either a recommendation or endorsement by English Heritage
or its agents.

Accuracy of information
While every effort has been made to ensure faithful reproduction of
the original or amended text from authors in this volume, English
Heritage and the publisher accept no responsibility for the accuracy of
the data produced in or omitted from this publication.

Front cover
William Martin, the Rural Industries Bureau's South West Region
thatching officer instructing master thatchers, 1949
(Rural History Centre, University of Reading).

Contents

Acknowledgements and Authors	vi
Foreword	vii
Preface	ix

PART I: THATCHING PLANTS AND THEIR USE — 1

Chapter 1 Introduction to the three common thatches today — 3
- Long straw — 3
- Water reed — 5
- Combed wheat reed — 8
- Distribution of the three common thatches — 9

Chapter 2 Water reed and other aquatic materials, by John Letts — 10
- Distribution of sources of water reed — 10
- Quality and decay — 13
- Ecological issues — 14
- Other aquatic materials used for thatching — 14

Chapter 3 Thatching straw, by John Letts — 16
- Botany — 16
- Decay — 16
- The assessment of straw quality — 16
- Plant breeding before 1940 — 17
- Improvements in grain quality — 18
- The origin of Plant Breeders' Rights legislation — 18
- Developments during and after the Second World War — 19
- The adoption of Plant Breeders' Rights and the National List — 20
- The National Institute of Agricultural Botany — 21
- Recommended List — 21
- Varieties used for thatch — 21
- Rye and *Triticale* — 22
- Growing thatching straw in England — 22
- Conclusion — 24

PART II: ORGANISATION AND TRAINING — 27

Chapter 4 The Rural Industries Bureau and its successors 1940–94 — 29
- The purpose of the Rural Industries Bureau — 29
- Thatching and the Bureau — 29
- The impact of the Second World War — 30
- Redefining thatching — 31
- Materials according to the Rural Industries Bureau — 33
- The Council for Small Industries in Rural Areas and The Rural Development Commission — 49

Chapter 5 Trade organisations — 51
- The Master Thatchers' Associations — 51
- The evolution of national trade organisations — 53
- The National Society of Master Thatchers' Associations — 53
- The National Society of Master Thatchers — 54
- The National Council of Master Thatchers' Associations — 54

Chapter 6 Training, 1940s–94 — 56
- Training before 1940 — 56
- The influence of government organisations on training since 1940 — 56
- Training in 1994 — 58

PART III: CHANGES OUTSIDE THE THATCHING INDUSTRY — 59

Chapter 7 Changes in production and supply of thatching materials — 61
- Straw: changes in farming from 1940 — 61
- Straw production in the 1990s — 63
- Changes in water reed production, 1940–90s — 64

Chapter 8 Changes in demand for thatch — 66
- New owners, new status — 66
- Thatchers and owners — 70

PART IV: CONSERVATION AND PLANNING — 73

Chapter 9 Conservation issues — 75

PART V: APPENDICES — 81

Appendix 1 Thatching survey, 1965 — 83
Appendix 2 A trainee thatcher reflects on her trade in 1950 — 87
Appendix 3 Historiography — 89

Glossary — 93
Bibliography — 95

Acknowledgements

This book would not have been produced without the help of numerous individuals: thatchers, house owners, conservation officers, historians of all descriptions, archivists, researchers, members of staff of English Heritage and members of staff, past and present, of the Council of Small Industries in Rural Areas and the Rural Development Commission.

I am most heavily indebted to John Letts and Dr James Moir, whose research into thatch for English Heritage began before mine (Moir & Letts 1999). Their knowledge, advice, comments and views have been crucial to the production of this book. They introduced me to Peter Brockett, a thatching instructor and independent thatch consultant who was dedicated to all aspects of thatching and related research. His early death is major loss to thatching. Peter's advice and comments on the early drafts on which this text is based have been invaluable. David Brock and Nicholas Molyneux of English Heritage provided encouragement and advice throughout. Dr Kate Macdonald edited the text with patience and tact.

Among many thatchers, working or retired, who took the time to write or talk to me and show me thatched buildings or commented on various drafts of the text and to whom thanks is due, I would particularly like to mention the Death family, Jack Dodson, Alan Fooks, Tristan Johnson, Jeff King, Alan Lewis, Rod Miller, Sid Pearce, Alan Prince, Jem Raison, David Trezise, Bob and Alistair West, Chris White, Derek Wisbey and Harold Wright. The National Society of Master Thatchers, the National Council of Master Thatchers' Associations and regional Master Thatchers' Associations were unfailingly helpful and I am especially grateful to the Somerset Association, who preserved and allowed me to see the minutes of their past meetings.

I would also like to thank many conservation officers, in both county and district councils, for talking to me about thatch and thatching policies, particularly Michel Kerrou (Northamptonshire), John Lowe (Dorset), Peter Child (Devon), Beth Davies (South Cambridgeshire) and Ollie Chapman (Kent).

I would like to thank staff at the Public Record Office and the former Rural Development Commission who, between them, managed to arrange for me to see documents. Staff at the Rural History Centre, the University of Reading, the RIBA library and various local studies libraries and record offices have also helped to make this volume of the Research Transactions possible.

Building historians, particularly members of the Vernacular Architecture Group, gave good advice and information: particularly Malcolm Airs, Peter Smith, Michael Laithwaite, Ray Harrison, and Dr Ellen Van Olst who gave me an introduction to Rijkdienst voor de Monumentenzorg, who answered all my questions and gave me access to their library. Geoff Baker translated portions of *Het Weke Dak* (1990). Gordon Glover's knowledge of agricultural machinery in the south west was invaluable.

Finally I would like to thank John Thorp, my partner in Keystone Historic Buildings Consultants, for unflagging interest and advice, as well as our associates: Dr Anita Travers for investigating some of the primary sources and Rupert Ford and Sophie Sharif for help in preparing illustrative material.

Jo Cox
Keystone Historic Buildings Consultants
3 Colleton Crescent
Exeter
Devon EX2 4DG

All sources for Figures and data for Tables are given in brackets in their captions. Please refer to the Bibliography for textual souces. All sources, originators and/or copyright holders are given where known. Every effort has been made to trace copyright holders, but English Heritage would be glad to receive any information updating the present list.

AUTHORS

Jo Cox is a partner in Keystone Historic Buildings Consultants which she established with John Thorp in 1987. The practice specialises in combining documentary research, analysis and recording on buildings of all dates and types. Jo is a graduate and post-graduate of the University of Exeter and now an Honorary University Fellow in the School of English.

John Letts is an archaeobotanist at the Department of Agricultural Botany at the University of Reading, and is the author of *Smoke-Blackened Thatch* (Letts 1999).

Foreword

This volume, and the previous one in the Research Transactions series, *Thatching in England 1790–1940* (Moir & Letts 1999), hold the results of two research projects commissioned by English Heritage in 1993 to establish the history of thatch and thatching in England. The research underlying this volume was the subject of a tight brief, and covers a period which is much more difficult to elucidate with historical objectivity, due, in part, to the lack of primary source material deposited in public archives for the period after 1950. Nevertheless, it has been possible to lay out the history of this period with a great degree of clarity even for the very recent past.

The stimulus for the work presented in these volumes resulted from concern surrounding the future of thatching as an industry, trade, craft or art in England. These concerns form the main part of the story elucidated by Jo Cox's research, published in this volume. The combination of pressures upon thatchers from the building conservation profession, most notably in those areas where there were planning conflicts between the use of water reed and wheat straw, as well as the use of combed wheat reed versus long straw, was also significant.

As a result English Heritage decided to commission research into the recent history of thatch. Whereas the work on smoke-blackened thatch (Letts 1999) was intended primarily as a study in archaeobotany, because of the great interest in the physical survival of early plant material, the roots of the two *Thatching in England* volumes in the English Heritage Research Transactions series lay in a practical desire to understand a history which has been unusually clouded by myth and controversy. In few occupations can there be so much consciousness of tradition, but so little firm information about past techniques and materials. The study was therefore to be one in which a narrative would be paramount, but the account should also rely, wherever possible, on archaeological evidence. It was undertaken in two parts, breaking the story at 1940; covering the agricultural revolution in the first study (1790–1940) and the period of living memory (1940–94) in the second.

The work revealed that there was a much more complex picture than had previously been recognised, and that this apparently ephemeral material has not only left extensive traces in the archives, but also survives in substantial quantities on roofs from the medieval period onwards. This has emphasised the importance of undertaking the archaeological research alongside the historical research.

There is an obvious chronological gap between Letts (1999) and these two volumes, in which there is much material still to be studied. The researchers have only been able to scratch the surface of some of the sources, so further studies on a national and regional scale are still much needed. Two examples of lacunae identified by Dr Cox since writing this volume are that there are no technical details here on the question of fitting of thatch on roofs and how this aspect of thatching has changed in the post-1940 period. However, in view of the considerable local variations this is something that will need to be studied on a detailed regional basis. A general topic that has not been researched in any depth and would repay further work is the complex relationship between thatchers and insurers in the period, which is a large subject in its own right.

Longevity of materials has long been one of the major themes of writers on thatch, and it is clear that this would merit more research. Letts & Moir (in volume 5 in the series) have shown the extent to which this question relates to the length of the straw, but we know that the factors are so varied and complex and make scientific quantification very difficult. Examples from the three main modern thatching techniques have survived for more than fifty years, and roofs with smoke-blackened thatch have lasted for at least 400 years, so this can hardly be thought of as a short-lived material, particularly when the comparative longevity of non-traditional modern roofing techniques are considered.

The emphasis of the research has been on water reed and wheat straw, the major surviving materials on roofs. This has led to an inevitable southern focus in the work since this is where the vast preponderance of surviving thatched buildings are located in England. However, the research does cover all the materials known to have been used for thatching English roofs.

The content of this volume stops at 1994, the year in which the research was delivered. The contents thus do not cover the most recent developments within the subject. There have been some significant pieces of scientific research and practical guidance on thatch and thatching carried out since 1994: the Building Research Establishment, in collaboration with the late Peter Brockett (to whom all thatching historians owe a tremendous debt), looked at the issue of accelerated south slope decay; the work of RHM Technology on fire prevention (Angold *et al* 1998) and their preliminary work on factors affecting longevity

(Angold & Sanders 1998); the work of West Dorset District Council in their flexible application of the Building Regulations to accommodate new thatched buildings and the BRE work looked at ventilating new thatch. There has also been recent research on the historical and archaeological aspects of thatch and thatching in Ireland (by Letts) and the work in Scotland reported in two recent publications (Walker *et al* 1996; Holden 1998).

The National Vocational Qualification has now been developed as far as level 2, and level 3 will be launched shortly. The Rural Development Commission, which with its predecessor the Rural Industries Bureau, subsequently the Council for Small Industries in Rural Areas, plays such a central role in the post-1940 story of thatching, was transmuted into the Countryside Agency in 1999. It has committed itself to the continuance of its most vital asset, the thatching training school at Knuston Hall, Northamptonshire.

A source which has become available since the research for this volume was undertaken is the computerized statutory lists of listed buildings. This provides us for the first time with an accurate national figure for the number of thatched buildings. In January 2000 24,059 were recorded. A map illustrating the current distribution shows a broad band across the country running from the deep south-west of Devon to East Anglia (Brock 1999, 31).

The acute (if localized) shortage of thatching straw in 1997 and 1998 as a result of two bad harvest years has led to the beginnings of a new trade, the import of straw from Poland, and the even more distant trade in South African grasses. English Heritage is now addressing this issue with some of the various responsible bodies. Progress must be made if we are to ensure that local sources continue to be used for this most sustainable of roofing materials.

At the time of writing English Heritage is also on the verge of a new round of detailed research, with an ambitious programme of research planned by the National Trust on their estate at Holnicote, on the edge of Exmoor. There will also be a further detailed county study, with Jo Cox's forthcoming volume on Devon thatching.

The research reported here, and the contacts established with the thatching industry, has formed the major part of the background to the production of English Heritage's Guidance Note *Thatch and Thatching* (English Heritage 2000). This seeks to provide a framework for the preservation of our historic thatch and thatching traditions within the overarching historic building legislation and guidance.

Nicholas A D Molyneux
Inspector of Historic Buildings
English Heritage
Bristol
January 2000

Preface

This volume of the English Heritage Research Transactions covers thatching in the modern period, concluding in 1994 when the dust of recent events and the sources available for research had scarcely settled down into history.

In the 1980s large numbers of thatched buildings were added to the list of buildings with statutory protection. This brought into prominence gaps in understanding, not only the ancient history and survival of thatch, but also the post-1940 inheritance of change, which is traced in this volume. This period has seen acute pressures on the availability of the plants employed for thatching; the development of formal systems of organisation and training for thatchers as their craft has also become an industry; and the shaping influences of external change: technological, demographic and social. All these themes are active in the current debate about the place of thatch in building conservation.

The research materials for this volume were diverse. Researchers of recent history are obliged to work in the time-lag between primary documentation remaining with its originator, and the process of weeding and depositing material as 'archives'. Fortunately this is a gap that can be bridged by living people, provided that it is understood that the difference between information and opinion is more sensitive in oral history than when written records are used. Most of the relevant primary documentation from the series of government agencies who were directly involved with thatching from 1940 was accessible for this volume. This proved enormously valuable, particularly the records of the Rural Industries Bureau. These represent not only the first centralized perspective on thatching, but a perspective written down by or based on the understanding of experienced thatchers. This made them not only worth investigating, but quoting extensively.

The main text is organised into four parts. The critical nature of the quality and availability of suitable materials is a persistent theme throughout. Materials were, therefore, the natural starting point for the text. Part I is a comprehensive background to the use of water reed, other aquatic materials and straw for the three common thatches in the period. John Letts provides a thorough understanding of these materials in Chapters 2 and 3. These cover their botany, how they decay, how and from where they have been sourced and how they have been evaluated for thatching.

Part II turns to consider the development of house-thatching into a self-conscious industry. The role of successive government agencies has been significant to this development. One of their legacies was a contribution to the shift of combed wheat reed and, later, to water reed into what had traditionally been long straw areas. The agencies built solid foundations for the industry of the 1990s. They encouraged first regional and then national trade organisations. They provided training schemes responsive to the needs of thatchers. They were pioneers of research into thatching problems. These were solid foundations on which the industry of the 1990s was built. On the other hand, perceptions of the relative merits of the three common thatches, based on the evidence of the quality of materials and the skills of thatchers in the 1950s, have never been left entirely in the past.

Part III examines change outside the industry that has directly affected thatching since 1940. The supply of materials has been affected by harvesting techniques. Mechanization benefited water reed harvesting but was one element in the split between thatching and arable farming. This led not only to the production of thatching straw as a specialist crop, but created shortages that, from the 1950s, opened up a vigorous trade in imported water reed. Demographic change from the 1960s brought new owners and investment to rural properties. This continued to sustain the thatching industry in 1994 but at the expense of worthwhile elements of earlier practice, notably re-dressing or patching to extend the life of a thatched roof. The erosion of a rural culture familiar with thatching placed the thatcher in a position of sole authority on thatching matters.

The building conservation debate in 1994, generated by the numbers of thatched buildings listed in the 1980s, is the subject of Part IV. It is hoped that this volume, commissioned as a result of that debate, provides material that makes a useful contribution to it. Part IV examines the relationship of thatch, thatchers and thatching to the concerns of building conservation, specifically to listed building legislation.

The Rural Industries Bureau 1965 Thatching Survey has been reproduced in full as Appendix 1 as an exemplary and concise account of the industry and the perceived pressures on it at that date.

One of the many surprises in the documentation was a 1950 letter by a woman thatcher whose wry observations on the training she was given are reproduced in Appendix 2. Secondary sources proved interesting enough to warrant an appendix on historiography (Appendix 3).

Part I
Thatching plants and their use

1 Introduction to the three common thatches today

There are three principal thatches today: long straw, water reed and combed wheat reed. Although statistical information is meagre, most thatchers today are capable of fitting at least two of the principal thatches, whereas their predecessors in 1940 were only likely to have been able to fit one. The 1994 training course, the New Entrants Training Scheme, leading to a City and Guilds qualification, required competence in two.

Some understanding of the difference between the three thatches is critical to an understanding of post-1940 changes. It is not easy to define any of the three thatches with authority. Neither materials nor techniques are standardized. Superficial appearance varies a good deal, especially for long straw. As far as fitting goes, distinguishing between what is convention, and what is necessity for effective performance, is not easy. Thatchers themselves will not all agree on this matter. Practices considered to be unorthodox in some areas are conventional in others. A number of thatchers use certain techniques because 'that is how it is done' and may not recognise that there is a performance reason behind the technique. The classic example of this (Paul Norman, pers comm), is the butts-up ridge. Here the join in the ridge, for performance reasons, should be placed on the non-weather side of the building. Some thatchers, who are perfectly skilled technically, may place it facing the weather side, creating an unnecessarily weak point. Others, more thoughtful, may use an unorthodox technique because they understand that it will get the best out of the materials they are using. It is said, with some justice, that a really skilled thinking thatcher can do just about anything with any material, if he is so minded, and could make a reasonable fist of thatching with anything from bracken to bamboo, adjusting technique to get the best out of the material.

LONG STRAW

Long straw (Fig 1) is now usually thatched with wheat straw, from a limited number of available varieties of wheat that produce a straw long enough to be suitable for

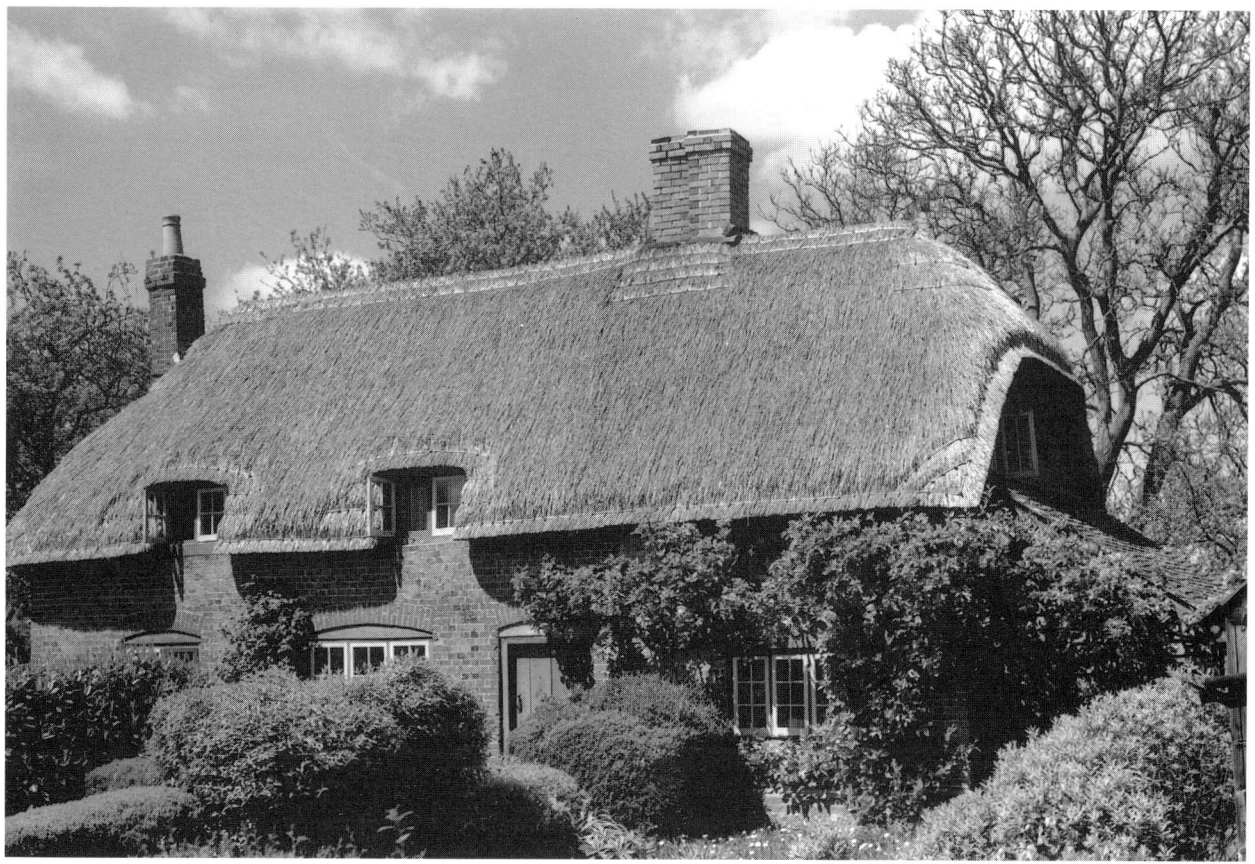

Figure 1 Rowlands Castle, Hampshire, 1950s, thatched with long straw and a half hip (Victor Schafer/Rural Development Commission).

thatching. Rye was also commonly used before the Second World War and straw from other cereals has been used in the past. *Triticale,* a rye/wheat cross, is sometimes used today. Long straw consists of stems of threshed wheat, usually about 750–900 mm long on average. The stems will have been softened or crushed to various degrees, by the action of the threshing machine.

Before 1940 straw for long straw thatching was produced on farms as a by-product of cereal production. Thatching was one of the uses to which straw was put in an agricultural context. It was used as the predecessor of the tarpaulin, to keep rain off stacks, root clamps, wood piles, even farm machinery that was not needed until next season, and farm buildings. House thatching also made use of straw which, in a rural context, was free or waste material. Long straw thatching, like combed wheat reed thatching, used to be the Siamese twin of English mixed or arable farming.

Changes in agriculture since 1940 have created an apartheid between the production of grain and the production of straw for long straw thatching. Wheat breeding, the use of artificial fertilisers and the development of harvesting machinery (discussed in detail in later chapters), have all contributed to a sea change in the supply of thatching straw. In most cases this is now grown as a specialist crop. The quality of straw available for long straw thatch is said to have improved since a low ebb in the late 1960s and 1970s and the general quality of what can be obtained today has risen since. This is a judgement based on thatching experience in the period, made in an eloquent, but unpublished, draft article by Peter Brockett, 'Straw thatching: when is a straw not a straw?' (private collection). However, older thatchers claim that it is still not as durable as it was before the 1940s. Whether or not this is a trick of memory, the central problem with long straw today is assured supplies of good quality material. The long-term future for long straw depends on the maintenance and development of what are specialist supplies and the maintenance, reintroduction or breeding of suitable varieties of wheat for producing thatching straw.

Method of supply

Before 1940, thatchers could obtain straw directly from local farms, on which they often worked, or through local straw agents. Today the straw may still be obtained directly from a farm with the necessary (old-fashioned) equipment to produce it as a specialist crop, or using the modern header stripper method of harvesting, although some thatchers are doubtful that this method produces authentic long straw. Long straw cannot be produced when a combine harvester is used. A straw agent may be a farmer, or another thatcher, or the thatcher may rent or own enough land to produce the material himself. Straw for thatching was not imported from foreign sources in 1994 but has been imported since. Long straw thatchers in Wiltshire, for instance, use straw grown both in Cornwall (where there is no long straw thatch put on roofs) and in Essex, as well as material produced locally (Alan Lewis and Sid Pearce, pers comm).

Fitting

Long straw is more difficult to define than either of the other thatches because the term covers a greater range of practice on the part of thatchers today, as well as a far greater range in methods of production and practice in the past. Practitioners themselves find a definition of long straw difficult and the account given below is deliberately generalised.

Like combed wheat reed, re-thatching in long straw is usually an overcoat onto existing straw. This means that only enough old straw to reveal a sound base is stripped back and the new thatch is fixed, not directly to the roof timbers, but into old straw. Historic layers of straw and other archaeological evidence are thus likely to be preserved below the new thatch. The oldest layers of straw, against the roof carpentry, can be medieval.

The principal difference between long straw and the two other common thatches is that the material is not dressed into position with a tool, as combed wheat reed and water reed are, but is bedded onto the roof. Long straw also traditionally requires more preparation on the ground by the thatcher than either combed wheat reed or water reed, the straw being 'yealmed' or drawn into compact layers prior to fitting on the roof. The method of bedding, rather than dressing the straw up with a tool, means that longer lengths of the material are initially exposed on the surface by comparison with the other thatches. Exposed lengths of straw usually include the threshed ears of wheat (which may appear in various different proportions to the butt ends of the straw), giving the material a rougher texture compared with what has been called the 'cropped' or 'quill-like' appearance of water reed and combed wheat reed (Brockett & Wright 1986, 3). The degree of roughness of texture varies considerably and depends in part on whether the surface is clipped or not, although the surface wears away in two years or so, revealing new ears. Dormer windows, gables and verges may be treated rather differently in long straw from the other two thatches. External hazel rodding, fixing the thatch, can be seen at the eaves and the gables on a long straw roof, but is rarely seen on the other two thatches. The threshed ears, which can contain the odd grain of wheat as well as insects, may attract birds, and, since the Second World War, most long straw roofs have been covered with wire netting to prevent bird damage.

Advantages and disadvantages for a thatcher

It is argued in some quarters (and disputed in others) that long straw is the most skilful of the thatching techniques, taking longer to learn than the others. Preparation of the straw on the ground requires particular skill. Thatching instructors reiterate that it is far more difficult to cross-train a thatcher familiar with combed wheat reed and/or water reed in long straw, than vice versa. Yealming, the preparation work, is laborious and some regard it as back-breaking. In recent years, these peculiarities of long straw, and its increasing rarity in some areas, have made it a particularly attractive material to self-defined craftsmen thatchers and to small-scale firms, who may find that there is less competition in pricing up for long straw, because in some areas few thatchers are skilled in fitting it and the larger firms may see it as less profitable. In 1965, by which time most thatchers were capable of fitting more than one of the three thatches, more English thatchers were capable of fitting long straw roofs than either of the others; 74%, compared with 55% for combed wheat reed and 37% for water reed (Rural Industries Bureau Census of Thatchers, private archive). Today far fewer thatchers are skilled in long straw work, allowing thatchers in some areas to see the use of long straw as a niche market. Its advantages to some thatchers, ironically, represent disadvantages to others, who may prefer a thatch that does not require laborious preparation on the ground and is quicker to fit.

There are drawbacks, even to thatchers who advocate long straw, drawbacks shared with combed wheat reed thatch. Because its production is specialised, some long straw thatchers who are closely involved with the growing and harvesting of the material find themselves preoccupied with the business of producing material.

Prices may fluctuate as the result of a poor growing season for straw, which brings scarcity. The price of the material tends to rise just before harvest, as last year's supplies become scarce. Fluctuations in price make it more difficult for a thatcher with a waiting list to price up for materials in advance than with water reed. Thatchers can avoid some of the problems of supply by growing their own straw on rented land, treating the grain, rather than the straw, as a by-product. East Anglian thatchers showed a preference for this from the 1950s and some thatching firms are as committed in time and energy to the production of their own material as they are to putting it on a roof. Long straw is bulkier to transport than either combed wheat reed or water reed, both of which are presented to the thatcher in tied bundles. A thatcher specialising in the production of long straw, or supplying it to other thatchers, will have good commercial reasons for preferring long straw thatch.

The customer

Long straw has suffered from a bad press since 1940 relative to combed wheat reed or water reed. Its reputation from 1940 up to the 1980s was of being a less durable material than either of the other two thatches. This view was given the credence of officialdom by the government agencies preoccupied with thatch, for a complex variety of reasons, discussed in detail in Chapter 4.

In the 1950s the government agency with an interest in thatching, the Rural Industries Bureau, believed that wheat-breeding programmes would mean that long straw would cease to be available, that the material was less visually attractive than the other two thatches, that long straw thatchers were less competent than other thatchers and would be better re-trained as combed wheat reed thatchers and that the material was far less durable than either combed wheat reed or water reed.

The reputation for the poor durability of long straw occurred in precisely the period when the question of durability came to the fore, as a new type of non-agricultural owner saw this as the crucial factor in the choice of materials. The figures for the durability of long straw, published in the Rural Industries Bureau's book *The Thatcher's Craft* (Morgan & Cooper 1960), and its second edition produced by the Council for Small Industries in Rural Areas in 1977, are still used as reference points today. These figures were hedged round with qualifications.

> We would like to point out that this [the length of life of the various thatching materials] cannot be given with any accuracy as it is dependent on so many factors, for example quality of crop and materials, weather conditions, situation with regard to prevailing winds and trees, of considerable importance whether or not a skilled thatcher is employed. (Morgan & Cooper 1960, vi)

When the figures are quoted today these important qualifications are not always quoted. The figures still play a significant role in the perception of long straw (and all the other thatches) by an owner. As with the figures for the durability of other thatches they are in need of revision, based on a regional assessment and taking the question of thatching standards into consideration.

Some owners may be sceptical about the published figures for the durability of long straw or swayed by the judgement of a thatcher that a long straw roof will last considerably longer than the figures quoted in *The Thatcher's Craft*. They may be keen to use long straw on the basis that it is the 'proper' thatch for their vernacular house, because it is or has been the thatch on their house and is commonplace in their locality. The rougher appearance of long straw, often described as 'shaggy' by comparison to the 'clipped' appearance of other thatches, appeals to the taste of some customers for authentic vernacular finishes.

Owners may have mixed feelings about the fact that it is not only possible, but desirable to patch long straw, thus keeping a roof going over a longer period than if it is neglected. 'Maintenance-free' has become a catchword for desirable building products.

WATER REED

Water reed (*Phragmites australis*), cut from managed reed beds (Fig 2), provides a thatching material for which there is evidence of use from medieval times. It is a good deal longer than the straw available for thatching in 1994, 1m to 2.5m, with tapering stems. Indigenous supplies used in English thatching, sourced from the Norfolk Broads and from other local water reed beds, have been overtaken by imported water reed from sources including Hungary, Turkey, Austria and France, although Norfolk is still a major source of supply, producing rather less than one quarter of the estimated gross aggregate demand (Environmental Appraisal Group 1991, 62). Importation began in earnest *c* 1950, although there is evidence of small-scale imports for thatching before the Second World War. Imports rose between 1979 and 1987 (see Fig 28) and are probably continuing to rise.

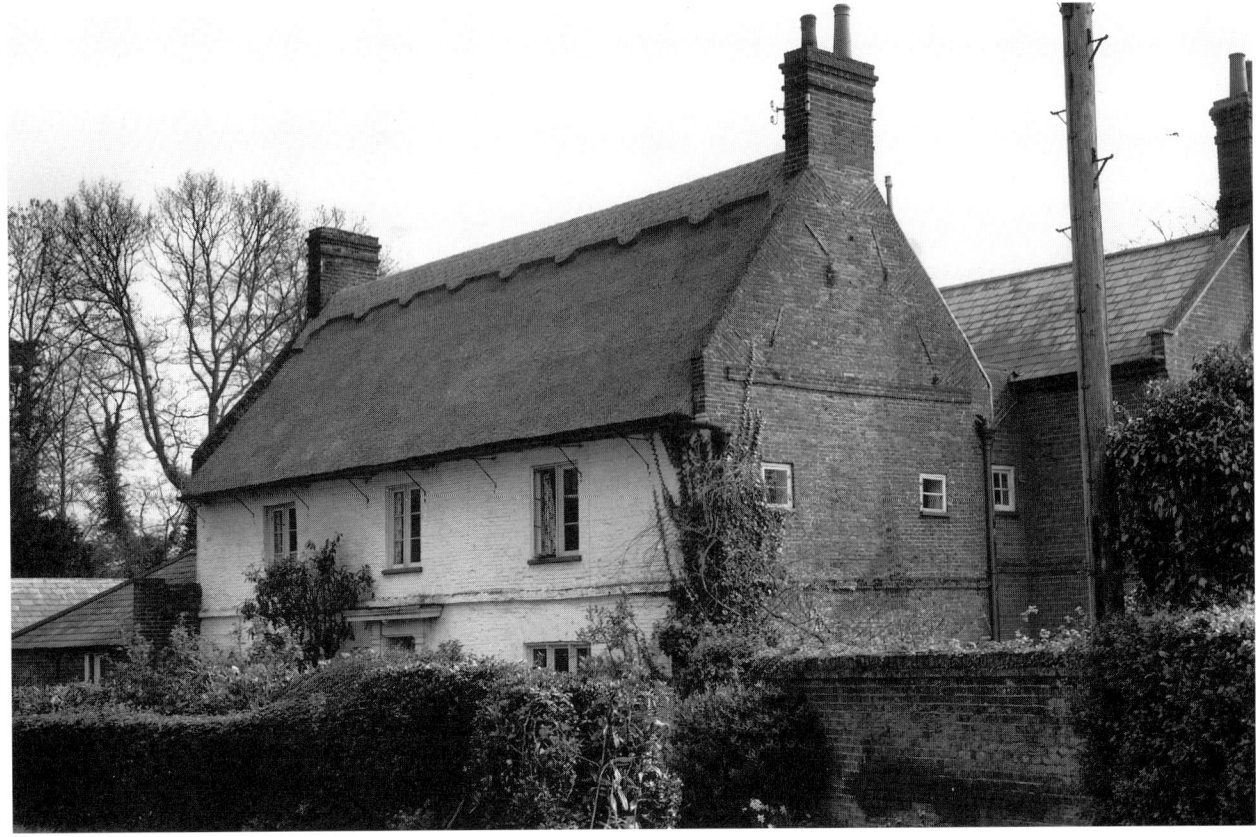

Figure 2 House thatched with Norfolk reed, 1959 (Victor Schafer/Rural Development Commission).

Method of supply

Water reed is usually sold to thatchers, or to agents, by the growers or cutters. The way in which this system operates varies from place to place but the material has a far longer history as a commodity sold and transported through middlemen than either long straw or combed wheat reed. In the post-war period when reed-cutters were in short supply, some thatchers rented reed beds from the owners and cut their own reed (J and B Death, pers comm), but harvesting is commonly undertaken by specialist reed-cutters. Very few thatchers cut their own reed, and some small-scale reed beds that were used after 1940, in Suffolk and Dorset for example, are no longer harvested for thatch. Today most English reed-growers sell direct to thatchers (Peter Brockett, pers comm).

After harvesting, imported water reed may pass through the hands of large European agents, a small number of whom have large-scale businesses, organising the purchase, transport and selling-on of reed, which is used in several European countries with a thatching tradition. These agents may sell-on to English agents (who may be thatchers themselves). Reed may be marketed by word of mouth or by sending advertising material to working thatchers.

There are drawbacks to the supply of imported water reed. It is subject to the vagaries of political change (supplies from Poland and Romania have been affected by political upheaval) and to changes in the exchange rate. If the reed that turns up in a container is not of the quality expected by the thatcher, it may be complicated to return it to a agent in another country, and there is a temptation to use it anyway (rather than delay a thatching job), or pass it on at a cut price. Thatchers using imported water reed have less control over the quality of the material than a thatcher who, say, is closely involved with the growing and production of straw, or the thatcher who goes to the reed beds in Norfolk to select the reed he or she wants.

Supplies of Norfolk reed have improved since the 1950s as a result of the activities of the Norfolk Reed Growers Association and the Norfolk Broads Authority, with encouragement from the series of government agencies who became involved with the thatching industry. At present, there is insufficient native water reed cut to support the demand from thatching.

Fitting

Water reed is fitted as thatch using a tool, a leggett (see Fig 34, top right), to dress the material, including the eaves, into position. This produces what has been described as the 'quill-like' or 'cropped' finish on a water reed roof, an appearance it shares with combed wheat reed. As a long-stemmed material, relative to obtainable straw, and dressed into position with a tool, water reed tends to produce a more angular thatch and outline than either long straw or combed wheat reed. Today it is usually associated with a block-cut sedge ridge when sedge is locally available, although straw can be used and was frequently used in the past.

Re-thatching an existing water reed roof with water reed often involves stripping the old water reed off the roof, back to the timbers. This method is also common when water reed replaces either of the straws, thus losing the archaeological evidence, both botanical and of old craft practice, that can exist in the lower layers of straw.

When water reed thatch is used today, it most commonly involves a fixing technique of driving metal crooks or hooks into the rafters. This requires better quality timbers than are needed when a straw roof is re-thatched, and a change from straw to water reed can result in replacement of historic timbers which would be adequate to carry a straw roof.

It is difficult to know how much convention and how much necessity is involved in the practice of stripping back water reed to the rafters. In Abbotsbury in Dorset, the local water reed, Abbotsbury spear, was traditionally overcoated onto earlier layers. This is a local practice which seems to be rare elsewhere in England. Perhaps the relatively slack pitches of the roofs close to the Abbotsbury reed beds and the comparative coarseness of the reed make the technique possible.

It should be said that the method of stripping back to the timbers when a straw thatch is replaced by water reed is not practised everywhere. The Cowell firm of Soham, exceptionally successful East Anglian thatchers at the turn of the century, considered it perfectly acceptable to spar-coat water reed onto straw roofs. For some years Devon thatchers, dealing with the relatively shallow pitched roofs in that county, have been spar-coating water reed onto existing coats of straw thatch. This technique is frowned upon today in some of the eastern areas. The East Anglian Master Thatchers' Association, for instance, states that it 'does not recognise the practice of "coating over" with Water Reed' (1989, 6), perhaps because the technique may be more suitable for the flatter-pitched roofs of the West Country than on the steeply-pitched roofs of the eastern region, where there is more risk of thatch slipping out of its fixings and off a roof.

There are similar issues of convention and necessity with regard to fixings. Screw-fixing methods using special drills can reduce potential damage to timbers and it is often possible to make use of methods that, although slower, (eg tying on water reed) are kinder to early roof carpentry than using crooks.

Advantages and disadvantages for a thatcher

There are a variety of different reasons why water reed is an attractive material for thatchers. It is quick to fit, relative either to combed wheat reed or to long straw. It is thus likely to be more profitable and opens up the possibility of covering more roofs per annum, and an increase in turnover. Arguably, it requires less skill to fit than long straw, and it is a technique which can be learned faster and is more conducive to working in time-efficient gangs.

Although water reed has its poor and good seasons, depending on the weather, the resultant price differences may be partly accepted as the agent's loss (Tristan Johnson, pers comm) and may not be passed on *in toto* to the thatcher or, because of the economies of scale, may be less significant than price differences resulting from a poor season for producing thatching straw. This is important to thatchers who have long waiting lists of customers, because it makes it possible to price up for a job more accurately, in the expectation that the cost of material will not change drastically from one season to another.

The scale on which water reed is produced for thatching in Europe means that shortages are less common than shortages in the supply of straw. If it is purchased through an agent, a continuous supply can be assured, whatever the time of year. In effect, by using water reed much of the angst, responsibility and expenditure of time, for a thatcher, associated with ensuring a steady supply of good quality straw for thatching with long straw or combed wheat reed, is shouldered by other individuals.

Thatchers who act as suppliers of water reed to others will have good commercial interests for preferring its use as a thatching material. This is a more pressing interest than thatchers/agents of straw, where supply is conducted in a national rather than an international context.

Some confidence in East Anglian water reed was lost in the 1970s due to reports of premature or 'accelerated' decay of water reed roofs. This encouraged some water reed thatchers, who had not already done so, to turn to foreign sources. It is more difficult to tell whether imported water reed may also suffer occasionally from premature decay. The sources are so diverse, and change so often, that reliable statistical information is particularly difficult to gather.

It should be plain from the above that there are a number of reasons of commerce and convenience, quite unrelated to conservation issues, why water reed thatching is particularly attractive to thatchers.

The customer

There are considerable attractions to water reed for an owner. Water reed has a reputation, well-deserved in East Anglia, of being the most durable of the thatches. From an owner's perspective, this means that the considerable investment in re-thatching is seen as giving the best return if water reed is used. Whether or not it is more cost-effective, however, depends on the relative price and durability of straw. Changing price differences between all the thatches make relative cost-effectiveness particularly difficult to judge.

The existing figures for durability of all the thatches, notably in *The Thatcher's Craft* (Morgan & Cooper 1960), are in need of revision, as indicated in the section on long straw above. The figures were produced when the bulk of water reed thatch was found in East Anglia, and have not been adjusted to take the spread of water reed since 1960 into account. It is unreasonable to suppose that a generalised figure for the durability of single-coat water reed thatch applied to a steeply-pitched East Anglian roof in a relatively dry climate will be the same as water reed applied over straw on a slack-pitched roof in the damp climate of the south west.

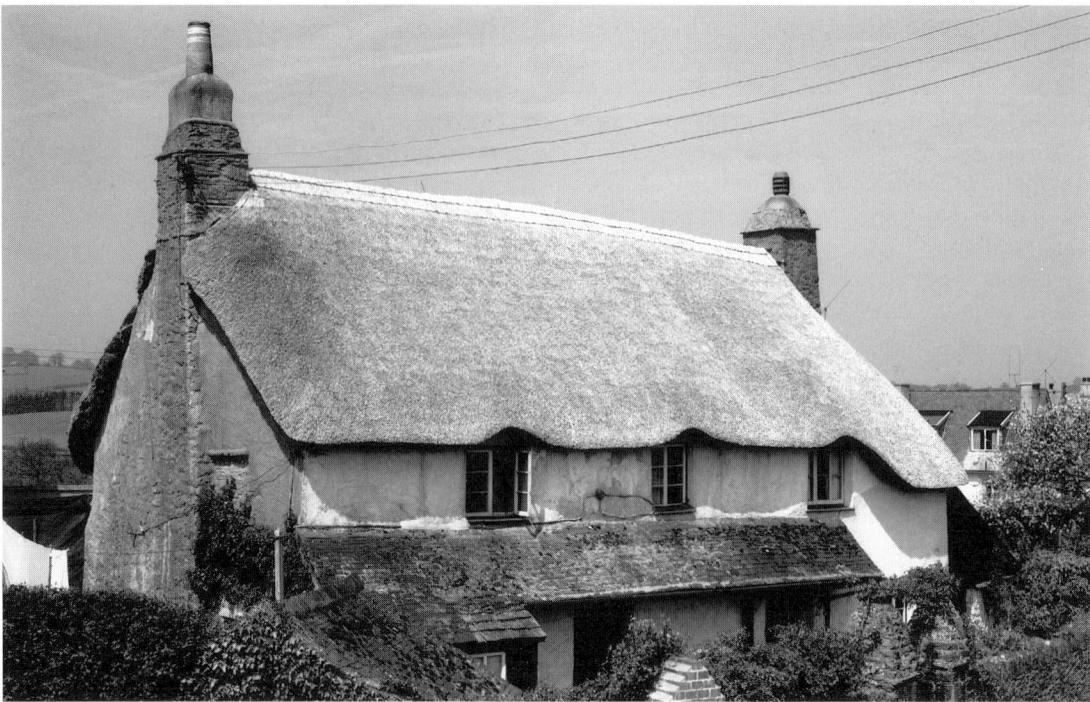

Figure 3 Combed wheat reed on a house in Dartington, Devon, 1959 (Victor Shafer/Rural Development Commission).

From an owner's perspective the mess and inconvenience of re-thatching may take place over a shorter span of time with water reed thatching, because it is quicker to fit onto a roof. A potential downside for a customer is that, since the quickest fit with water reed is onto new, sound timbers, there may be an impetus to extend the act of thatching into virtual re-roofing. It is also rare to patch water reed. This may not be seen as a disadvantage to an owner, who may view it as a more 'complete' thatch because, apart from re-ridging, maintenance is less likely. In fact there are methods of getting a longer life out of water reed, by drawing the existing reed down from the eaves upwards, adding some new material and re-dressing the roof, but these have largely fallen into disuse.

COMBED WHEAT REED

The name of this thatch is confusing because the material is actually straw, usually from the same varieties of wheat that are used for long straw. The 'reed' tag is connected to the method of fitting, which is similar to that of water reed. Combed wheat reed (Fig 3) thus occupies an odd position relative to the other two materials. On the one hand it is a straw thatch, on the other it has much more in common, visually, with water reed than with long straw. It must be categorised with long straw as regards the problems of supply, which it shares.

The production of straw for combed wheat reed is similar to that for long straw. The difference in production is that combed wheat reed is produced by removing the grain from the straw leaving the straw unsoftened by the threshing process and all lying in one direction, the heads at one end and the butts at the other. This process is carried out using a reed comber, an attachment which sits on top of an old-fashioned threshing machine, although straw using the modern header stripper method is also used. Hand combers preceded the mechanical reed comber. The reed comber requires additional labour, in what is already a labour-intensive procedure, using a threshing machine.

All the difficulties ascribed to the production of long straw above, apply to combed wheat reed.

Method of supply

The logistics of supply are similar to long straw although, for historical reasons, they are more likely to be produced by a farmer than a thatcher. Like long straw, combed wheat reed may be purchased through an agent or bought direct from the farmer. Some thatchers do grow the material themselves, but probably fewer, relative to the amount used, than grow their own straw for long straw. As with long straw, combed wheat may be produced close to where it is used, but may travel long distances. Tied into bundles, like water reed, it is easier to transport them than long straw. The bulk of combed wheat reed is still produced in the south west, though curiously enough, in Devon, its region of origin and where it was once ubiquitous, more (imported) water reed than combed wheat reed is used.

Fitting

Combed wheat reed is fitted, like water reed, by dressing it into position with a tool. However, unlike water reed, the eaves are cut, rather than dressed into position, and in most cases combed wheat reed is thatched on top of existing straw. This involves stripping off only as much old straw as is necessary to reveal a sound base into which the new thatch is fixed. As with long straw, the effect of this method, which has been in use at least since the medieval period, is to preserve many layers of historic straw below the new thatch, known as the coatwork. As with long straw this technique means that the bottom layer of thatch, against the rafters, may be medieval. Combed wheat reed coatwork can be fixed to a long straw base, although it is necessary to fix carefully as the coatwork is a stiffer material than the base.

In recent years some thatchers have fitted combed wheat reed long straw fashion. Rather than carrying out the laborious preparation on the ground for long straw, straw that has been put through a reed comber is fitted onto the roof but the surface is pulled about to a lesser or greater degree, in imitation of the surface appearance of long straw. External rodding is added at the eaves. This rodding, which is a necessity for fixing long straw, is largely ornamental when combed wheat in England is fitted to look like long straw, although it is routinely used with combed wheat reed in Ireland in windy areas to better secure the thatch. Some thatchers claim that fitting combed straw to look like long straw is just another long straw variant, or at least a perfectly acceptable substitute. Other thatchers and some conservation officers disagree, either on the basis of performance, or on grounds of authenticity, because the material is not yealmed.

Advantages and disadvantages to a thatcher

As far as fitting goes, combed wheat reed shares many of the advantages to a thatcher of water reed. Although less quick to fit than water reed, it is usually quicker to fit than long straw, and less labour-intensive once the straw has reached the thatcher.

In the 1970s some combed wheat roofs failed with surprising speed, especially in the south west. In some cases the thatchers concerned had used material that had every sign of being sound. Research carried out in the late 1980s at the University of Bath at the behest of the Rural Development Commission did not prove a scientific link between the rapid degradation of thatching straw and the use of nitrates during its period of growth (Kirby & Rayner 1986). However, it is generally accepted that the most successful husbandry of combed wheat reed and long straw involves only very limited use of nitrates. The Bath research encouraged some amendment of the technique of fitting combed wheat reed, demonstrating that a very tight and flat fit may jeopardise the performance of a roof by preventing the thatch from drying out quickly. The problem of accelerated decay had the effect of making thatchers (and owners) lose some confidence in the material and may have contributed to the shift from combed wheat reed to water reed in the south-west counties at that time.

The disadvantages to a thatcher relate to the problems of assured supplies of good quality material, which are similar to those experienced by thatchers using long straw.

A thatcher producing his own combed wheat reed, or supplying it to other thatchers, will have good commercial reasons for preferring it as a material.

The customer

Many owners find combed wheat reed attractive, often as the middle way in durability between long straw and water reed. Visually it can be very similar to fine water reed. It may be marketed to an owner as a poor man's water reed, sometimes less expensive, but with an assumed durability, once again based on the Council for Small Industries in Rural Areas figures published in 1960, greater than long straw. Whether or not the figures for the durability of combed wheat reed were influenced by its visual similarities to water reed has not yet been established, but the 'reed' rather than 'straw' tag in the name of this material may have had an effect on the way its durability is perceived by customers. The Council for Small Industries in Rural Areas's predecessor, the Rural Industries Bureau, was effective in promoting combed wheat reed eastwards from its origins in the south west, with claims for longevity that seem rather surprising now, but which may relate to the existence of better quality combed wheat reed being available in the 1950s than today.

In spite of its superficial similarity to water reed, there may be an aesthetic preference for combed wheat reed by a customer, for the same reasons of vernacular authenticity which may be attributed to long straw. Combed wheat reed gives opportunities for shaping a roof that are more difficult to achieve with the longer water reed. This can result in a fairly dumpy outline, which is rare in water reed. However, in some circumstances it can be difficult for a non-thatcher to distinguish a combed wheat reed from a water reed roof.

Owners may be put off by the poor publicity concerning accelerated decay which was noted in combed wheat reed in the 1970s and 1980s and described above.

DISTRIBUTION OF THE THREE COMMON THATCHES

The broad picture of the distribution of the three principal thatches in 1940 was that combed wheat reed was found in Cornwall, Devon, Somerset and probably parts of west Dorset, with no long straw thatch in Cornwall and Devon. Water reed was found principally close to the main sources of supply, the large managed beds in East Anglia, as well as in pockets close to local sources of supply, eg in Dorset near the Abbotsbury reed beds and, here and there, where it had been carried by water reed thatching firms from East Anglia who had travelled further from their sources of supply since before the war than straw thatchers commonly did. The boundary between combed wheat reed and long straw is difficult to identify with precision, but seems to have been in Dorset with combed wheat reed in the west of the county and long straw in the east. Outside the south west, versions of long straw were found in all the southern counties where thatch was still a relatively commonplace roofing material. This is a necessarily crude picture, and leaves out a number of sub-regional variations, including other materials, eg heather and sedge, used where these were readily available, but it is a useful starting point for understanding the changes to the distribution of the three common thatches from 1940. Regional thatching studies will, in time, help to identify the subtleties of each area, including diversity within the three common thatches both before and after 1940.

2 Water reed and other aquatic materials

JOHN LETTS

The common reed *Phragmites australis*, is a tall, stout, perennial monocot (see Glosssary) that is very common in wet, or at least permanently damp, soil throughout lowland Britain (Clapham et al 1987, 648). It dominates a variety of wetland habitats both inland and along the coast, and flourishes in all but the poorest and most acidic soils. It grows in the shallow fringes of many ponds, rivers and lakes, in sluggish ditches and streams, and anywhere that is wet, or at least permanently damp, year round. Reed requires at least 10 mm of water through the winter to late April, and will grow in up to 2 m of still water, but is less tolerant of fast-flowing water and extremely saline conditions. In ideal conditions, it forms large stands at the back of coastal salt marshes in river estuaries, and is particularly plentiful in the mires, marshes and fens of East Anglia and Somerset.

Reed spreads primarily by rhizomes, and can quickly invade and dominate a shallow water habitat to form a dense mono-dominant stand. The roots can penetrate anaerobic estuarine clay, and it can aggressively colonise field crops from adjacent drainage ditches on well-drained land. Only a small portion of the seed produced each year by a reed plant is viable, and only a small number of seeds become established because of its exceedingly narrow germination requirements, so that local populations are composed primarily of vegetative clones identical to the parent plant.

Natural reed beds are an ephemeral stage in the natural process of ecological succession from open water to dry land. Unless the dead reed is washed away, as occurs naturally in many coastal marshes, it accumulates to form peat and gradually raises the surface of the bed. Eventually, drier conditions in the raised bed allow dry land plants to colonise and reed is out-competed by scrub and eventually by woodland. Regular cutting removes the bulk of the annual production from a reed bed, and management allows almost pure beds to be maintained almost indefinitely. Many reed beds in the United Kingdom exist only because they were managed for centuries for thatch, and the existing beds are a remnant of what was once a much more valued and plentiful resource.

The best quality reed is cut every year (single wale), but most beds are cut every two years (double wale) to maximise the efficiency of the harvest. Beds are thus divided into sections that are cut in alternate years. Reed cutting was mechanized only in the late 1950s and 1960s. Cutting began after frost knocked the flag leaf off the dead stem in early winter (December), and had to be completed before spring growth began (in April). Reed bed management required the reversal of natural hydrological cycles: in natural beds, water levels are at their highest in the winter/spring and lowest in the summer. In managed reed beds water is drained away in the late autumn so that cutting can occur on a shallow or semi-hard surface in winter. The beds are then flooded in spring to stimulate growth, to protect the young shoots from late frost, to discourage competition from weeds through the summer, and to clear the beds of accumulating plant debris.

DISTRIBUTION OF SOURCES OF WATER REED

On strictly environmental grounds, much of the United Kingdom is suited to the growth of water reed (Rackham 1986b). According to the Institute of Terrestrial Ecology, approximately 1 million hectares of the United Kingdom is suitable for reed growing on ecological, hydrological and climatological grounds, but most of this land is prime agricultural land as in the Fens. One quarter of the United Kingdom is, or was once, some kind of wetland. Today there are only about 300 ha in East Anglia, 50 on the south coast, 150 or so on the Tay estuary in Scotland and smaller amounts in Wales. A 1982 survey revealed only 109 reed beds larger than 2 ha, with a total surface area of approximately 2,300 ha (Bibby & Lunn 1982). Many of these beds are recent in origin, and very few are inland. The largest single bed in Britain, in the Tay estuary in Scotland, is less than a century old. Approximately $\frac{1}{6}$ (400 ha) of this amount is managed for thatching.

In the past, particularly at periods when sea levels were higher and rainfall greater than at present, both inland and coastal reed swamp were much more plentiful, but there is very little direct evidence of it having been used for thatching. Even though it is very likely to have been widely used on prehistoric, Roman and Saxon roofs, very few assemblages of plant remains recovered from archaeological sites might reflect the use of reed as a roofing material. As Britain became an increasingly settled and agricultural society, wetland containing reed receded in the face of arable and pasture. The assarting of wetland increased most dramatically during the period of agricultural intensification beginning in the late Iron Age. Intensive cultivation on the light, fertile soils covering the gravel terraces adjoining the major river valleys of southern England led to drastic soil erosion, and to a massive infilling of river basins with clay alluvium (Robinson &

Lambrick 1984). This led to catastrophic winter flooding of the settled flood plains in the late Iron Age and Roman periods and to the spread of fringing wetland and reed swamp. This process continued in the later Saxon period and early medieval periods, but was greatly reduced as arable reverted to pasture after the plagues and depopulation of the fourteenth century. The predictable winter flooding of river flood plains gradually allowed them to be managed for animal feeding, eventually creating the botanically rich (but reed-poor) hay meadows that were the most valuable type of farmland in England until early this century.

Before the advent of effective sub-surface drainage in the early to mid nineteenth century, one quarter of the cultivated land in the United Kingdom was waterlogged for most of the year. Reed was undoubtedly plentiful in hollows, ditches and ponds throughout the country in the medieval period, but only a very determined thatcher would have been able to collect enough reed from small local sources to thatch a small house. Very rough calculations suggest that it would take six to seven years for a large farm, with access to ponds and miles of overgrown ditches, to collect enough reed to thatch a small house to half the thickness recommended today. Smaller farms in arable districts simply did not have access to sufficient reed to allow it to be used on a regular basis. Ample reed was available almost everywhere, however, for use as 'fleeking' (a thin lining of reed above the roof battens but below the thatch) and as a key for plastering walls and ceilings. In arable districts far from significant reed beds, thatching in reed was probably an impractical, and expensive, roof covering, particularly since adequate alternative materials such as straw, sedge, rush and heather were usually available.

Large wetlands with water reed persisted in some inland areas well into the nineteenth century, although not all carried significant amounts of reed. Four thousand acres (1619 ha) of bog survived at Otmoor (Oxfordshire) in 1800, but reed does not seem to have been used for thatching locally to a significant extent; and other major wetlands producing reed, sedge or rushes survived in Shropshire, Worcestershire, Leicestershire, Lincolnshire and along the Ouse River below Bedford. Gradually, however, wetlands were pushed back into their environmental strongholds in the Broads and the Fens of East Anglia, the Levels and the Moors of Somerset, the Lincolnshire Fens and the Humber Lowlands, the mosslands of Lancashire, the Romney Marshes, and the coastal marshes of the Thames estuary and a handful of tidal rivers along the south coast.

The Fens

The East Anglian Fens include over 1,200 square miles (3,108 km^2) of land that lie below sea level, but only a very small portion remains undrained (Grove 1962, Darby 1976). The Fens consist of seaward silt marsh and landward fen peat with a transition zone in between. The earliest peat accumulated in post-glacial times, and various stages of flooding, deposition of clay and return to wet conditions led to a further intermittent build-up of peat (much of it derived from water reed) dozens of feet thick. The natural vegetation of the Fens was probably a complexity of pools, reed beds, grassland, thickets and woods, depending on drainage conditions. The Fens were portrayed in early writings as inhospitable and unsettled (Rham 1845), but archaeological work indicates that they were well settled in prehistory and were the focus of drainage works by the Roman period. Anglo-Saxons settled on silt islets near the Wash and on islands of gravel and clay that rose above the peatlands, but settlement was sparse on the peat itself. Ad hoc drainage programmes continued in the medieval period, and with its ample resources of fertile arable soil and wetland, the Fens gradually became one of the richest regions in Britain. Numerous medieval records attest to the use of 'marsh resources' for thatching, and to the numerous laws and customs which governed their harvesting and sale.

Land was reclaimed from coastal salt marsh in the Fens by embankment along the coast, and by embankment of river systems to protect inland peatlands from flooding in the winter and spring. The history of draining in the Fens is complicated by centuries of unrecorded large- and small-scale efforts at reclamation. The overwhelming problem, which is a challenge today as much as it was in the past, is that the rivers running to the Wash drain a huge catchment and the flow can easily back up because the outflow must cross the higher embanked silt fens at the coast. In addition, peat shrinks when it is drained and cultivated due to simple breakdown of organic material (as in compost). Initially, the soil surface will drop by 1 foot (300 mm) or more a year, making drainage all the more impractical. The goal has forever been to improve the efficiency of the pumps that move water from low-lying fields up embankments into the drains and canals where it can be carried to the sea. Eventually, the peat decomposes, fully exposing the sands and clays which underlie the Fens and which are much less fertile and easy to cultivate.

Until the seventeenth century, drainage was piecemeal and reclaimed lands were frequently reflooded, but investors supported by landowners (and Parliamentary enclosures) hired Vermuyden, a Dutch engineer, to 'improve' the Fens through widespread drainage, in the face of local opposition. Fen dwellers were deprived of the common rights they had held for centuries, and were allotted small areas in the 'Poor Fens' in compensation. The success of Vermuyden's work was short-lived; the new drainage systems were often flawed in design, pumps were unreliable, and the best land shrunk away rapidly making drainage all the more difficult. Large drained areas eventually returned to wetland (generating new reed beds) or survived as wet, rough pasture. Pumping technology improved after 1820, and the high price of corn during the Napoleonic wars stimulated a final attempt at reclaiming what remained of the original fens. The last large open stretches of water were drained by the mid nineteenth century, facilitated by the use of steam pumps, and eventually by diesel and electric pumps. The Fens are still susceptible to flooding due to spring floods when high tides block the outflow to the Wash.

The Fen region is still one of the richest, and most intensively farmed, regions of the United Kingdom, but nearly half of the original peat has disappeared, and many rivers are 14 feet (4.27 m) or more above the surrounding farmland. Very little reed or sedge is available for thatching, and only small pockets of the original vegetation survive in nature reserves, such as at Wicken Fen. The agricultural 'improvers' of the nineteenth century believed wetland was a disgrace to a civilised agricultural society, and the onslaught on small pockets of wetland that have survived has continued, with official government support, into the present day.

The Levels

A similar story occurred in the Somerset Levels, a 200 square mile (518km^2) area of flat land that lies between the Mendips and the Quantocks (Storer 1985). This region was below sea level in 6000 BC, and had developed into a reed-rich estuarine swamp by 4500 BC. A build-up of silt and clay at the coast gradually formed a barrier against the sea, and a fresh water marsh developed dominated by sawgrass (*Cladium mariscus*) and sedge (*Carex* spp.). Peat accumulated, and an acidic bog cover had developed by the Roman period. Excavations of late Iron Age and early Roman period sites such as Glastonbury and the Mere lake villages have revealed settlements composed of roundhouses, and partially burnt layers of reed and rush have been interpreted by the excavators as the remains of thatching, but reeds were also probably used for fuel and the archaeobotanical evidence is far from clear (Bryony Coles, pers comm).

Traditionally, the Levels and the Moors provided summer grazing and peat for fuel, and were flooded in the winter and early spring. Records attest to the use of reed and rush beds for thatching in the medieval period. Significant drainage programmes in the Somerset Levels began in the Roman period, and large-scale Vermuyden-style endeavours had largely drained the Levels by 1850. Common lands were by this time enclosed and divided by rhynes (ditches) into easily managed and navigated rectangular units, and tidal sluices prevented reflooding by the sea. Drainage for peat extraction, and increasingly intensive management for agriculture, pushed the harvest of reed and sedge for thatching into obscurity. Now that large-scale commercial peat extraction is coming to an end, reed and sedge beds are once more becoming common. Reedmace (*Typha* sp.) is usually the first plant to recolonize abandoned peat workings, while sedges and rushes grow with water reed at the edges of ditches, particularly in older rhynes under traditional management. Rhyne reed is not grazed and is seldom cut for thatching. Although much of the peat in the Levels is derived from fen sedge, it has been extirpated from the area.

New reed beds have been created inadvertently in many parts of Britain over the past century as a result of industrial activity, construction for railways or due to the natural occlusion of coastal bays and estuaries by shifting sands, but little of this is cut for thatching. A small amount of reed was still being cut for thatching from rivers and ditches in the Oxford region in the mid-1970s, but the practice has been discontinued.

Reed production from local sources (ditches)

A large amount of reed is needed to cover a roof to modern standards, 1,500 bundles for a small cottage roof of 1,500ft^2 (139m^2). At least 500 bundles would be required to cover it to a reasonable degree. Since a reed bed managed for thatching can be expected to produce about 400 bundles of reed per acre (988 bundles per ha), a cottage could be expected to consume 1.2 acres of reed (or 0.5 ha). Even if five times more reed bed was available a century ago, it would still cover only 3,000 small cottages with a poor coat of 500 bundles. Without imports there was only enough reed to thatch 185 small cottages at the modern recommended depth of one bundle per square foot.

If ditch reed is taken into account other calculations show that not enough would be available for thatching. If a good managed reed bed can produce 400 bundles per acre (988 bundles per ha), dense un-managed ditches could rarely produce more than 200 (494 per ha), or they would not function as ditches. Perhaps 100 (247 per ha) could be expected from an overgrown ditch not recently cleaned, and only 50 (123 per ha) or less from reed fringing ditches. And not all of this reed would be of thatching quality. Most overgrown ditches probably yielded in the lower range. In a 3 yard (2.74 m) ditch, at very dense reed of 200 bundles per acre (494 per ha), one bundle could be collected every 8.8 yards (8.05 m), but only every 35 yards (32 m) at 50 bundles per acre (123 per ha). An average of 100 may thus be an exaggerated compromise, and could be collected every 17.6 yards (16.09 m).

Thus to thatch a cottage with 500 bundles would take 9.2 miles (14.8 km) of ditch at 50 per acre (123 per ha), at least 4.6 miles (7.4 km) at 100 bundles per acre (247 per ha) or 2.3 miles (3.7 km) at 200 per acre (494 per ha); a great deal of work. A large farm of 50 acres (20 ha), including pastures divided by ditches with two 1 acre (0.4 ha) ponds each, and 45 acres (18.2ha) of arable in three fields divided by ditches, would contain 4,107 yards (3755 m) of ditch and produce a maximum of 334 bundles at 100 per acre (or 167 at 50 acre, 825 and 413 per ha), but only half would be available for harvest due to neighbours' demands on these same ditches, labour costs, etc., so that perhaps only half of this amount (85 if at 50 per acre and 165 if at 100 per acre) would be actually available. A more typical 20 acre farm would similarly produce only 35 for 50 per acre and 90 at 100 per acre, insufficient to support a reed thatching industry. In actual fact, ditches were cleaned every four or five years and reed was not as common in medieval ditches as it can be in the modern day. Reed was also needed for fleeking (at least 30 bundles per roof), as a key for plastering ceilings and walls, and as fuel for baking and brewing.

According to these figures, the farmer of a 20 acre (8 ha) farm (a reasonable size in the medieval period) would have needed to collect water reed for 15 years in order to obtain enough reed from his property to thatch one small cottage with a thin coat (or 7 years on a 50 acre, 20 ha, farm). In arable districts, thatching with reed was possible only if local reed beds were available, or if other distant sources would be found. This was clearly possible on manorial estates, abbeys and monasteries with extensive land holdings. Reed, in the past, was of necessity an unusual roof covering in arable districts, and the poor had no choice but to thatch their roofs in straw, rush, heather or other local materials.

QUALITY AND DECAY

Premature decay was first reported in the 1950s, and became a significant problem in combed wheat reed roofs in the 1970s (see Chapter 9). Work at the University of Bath in the late 1980s on decay processes in combed wheat reed demonstrated that patches showing 'premature decay' were decomposing differently from the normal process of *Basidiomycete*-mediated (white rot fungi) decay that turns most organic matter into compost (Kirby & Rayner 1989). Straw from such patches contained cavities in their cell walls typical of *Ascomycete*-mediated (soft rot fungi) decay, particularly of the species *Haplographium*. It has been suggested that a similar process occurs in water reed roofs, but the evidence is not conclusive.

Research on the impact of nutrients such as nitrogen, phosphorous and potassium on the structure and strength of water reed has been underway for decades, in particular by Haslam in Cambridge and a group of German researchers (Haslam 1972). Young and Davies (1990) have recently demonstrated that variation in the thickness of sclerenchyma and parenchyma in reed stems was linked to habitat differences associated with farming practices, in particular the use of inorganic fertiliser.

The ecological, social and economic consequences of the abandonment of Norfolk reed by the thatching community would be significant, and concern elicited funding for a University of East Anglia project to examine reed quality and the impact of nutrient levels and management techniques on reed quality and decay (Bateman *et al* 1990). The goal of the project was to examine nutrient concentrations in reed beds and their effect on reed strength and rates of decay. Unfortunately, decay is extremely difficult to replicate in the laboratory, and no single feature of sediment chemistry could be used directly and conclusively to predict the rate of decay in stems. Stems certainly varied in their susceptibility to decay depending on the conditions in which they were grown, but how and why largely remains a mystery. Laboratory tests of decay are not necessarily reliable assessments of decay on a roof: for example, initial decay probably begins with fungal attack of soft parenchyma tissues within the core of the stem rather than the structural sclerenchyma, which gives the stem its strength. However, high levels of nitrogen in this soft tissue might stimulate microbial attack that would eventually attack adjacent woody tissues. No link was found between phosphorus and decay rates, although large amounts of phosphorus are thought to accelerate decomposition in most organic systems.

Although researchers have suggested that silicate opals (phytoliths) on cell walls confer mechanical strength and resistance to fungal colonisation, there is no evidence botanically to support this. They may simply occur as a result of silicates precipitating as a waste by-product of plant metabolism and have little structural role. No link was found between silicates and decay rates, although Haslam has found differences in the shape of phytoliths in different habitats, which are undoubtedly related to genetic differences between populations as well as environmental conditions. Nor was any direct relationship established between environmental iron, magnesium, sodium, calcium or acidity and reed decay rates.

The researchers did not find the expected link between nitrogen levels in dead stems and decay, but found enough evidence to justify further research. Nitrogen concentrations in dead stems were usually above the 0.2% nitrogen concentration that is believed would encourage predominance of a decomposer community of 'soft rot' rather than normal 'white rot' fungi, but since no patches of early decay have been noticed in water reed thatch, the researchers suggested that nitrogen in reed stems acts to speed up the normal succession of micro-organisms involved in natural decay of reed over the entire roof, rather than causing patchy soft rot as occurs on combed wheat reed roofs.

The results from the studies by the universities of East Anglia and Bath also demonstrated that variation in the decay rate of reed cut from different parts of the same marsh is as great as that from reed obtained from different marshes. Environmental concentrations of nutrients obviously vary greatly within and between marshes, and poor management (which leads to uneven drainage and heterogeneous growing conditions) probably accentuated differences in nutrient conditions both between and within reed beds. Reeds are excellent pollutant filters, and nitrogen-rich run-off will usually be cleared of its N before it penetrates to the core of a reed bed. Ambient nutrient levels will therefore vary in beds near sources of pollution and along rivers with heavy nutrient loads. Since most plants in a reed bed are clones of a common parent and are genetically identical, the great variations observed in the decay rate within one bed suggests that the susceptibility to decay is environmentally induced rather than genetic, but genetic variation must also exist.

The correlations between environmental conditions in reed beds with stem strength and durability in the universities of East Anglia and Bath research were tenuous, and complicated by the difficulty in measuring these aspects in the laboratory. Nor are such mechanical measures easily related to thatchers' assessments of quality characteristics such as strength, hardness and flexibility. Hard reed can be very brittle or very flexible, although it may perform equally on measures of tensile strength. Researchers with many years experience have abandoned mechanical methods in favour of controlled subjective comparisons using a reed of established quality as a norm, which more accurately reflects the subjective assessments of thatchers themselves.

The research by the universities of East Anglia and Bath demonstrated that dead stems did not reflect environmental concentrations of nutrients. Even though high nitrogen levels are thought to interfere with lignin production, the main thickening and strengthening material within plant cells, nitrogen concentration in dead stems had no bearing on strength tests in the laboratory, and stems varied in strength even though they had equal nitrogen concentrations. Nor was there a significant positive correlation between strength and concentrations of sodium, potassium or magnesium in sediments or water, even though potassium is a major influence on the synthesis of a plant's structural components. A little sodium (salt) may lead to a little toughening of the stem, but high sodium levels produce narrow brittle stems. Reed grown on acid peat was also relatively weak.

The strongest stems examined came from sites that were slightly brackish, where nitrogen in sediment was low, and where potassium was most plentiful. The authors postulated that it is the ratio of nitrogen to potassium that influences the sclerenchyma content, and thus the strength, of a stem. When the nitrogen to potassium ratio rises above 0.94, as it frequently does in reed beds flooded by waters drained from farmland with even moderate fertiliser use, the development of woody structural tissues is depressed relative to the production of soft parenchyma. This may be due to the fact that reed beds favour the retention of ammonium ions, which are the major source of uptake of nitrogen for a reed plant. Ammonium ions compete successfully with potassium ions for uptake sites within the reed, and thus the availability of potassium and lignin synthesis are depressed leading to weak stems. This retention of ammonium ions may be linked to uneven drainage and water flow caused by poor management.

An additional factor hitherto unrecognised in thatch quality and decay is epicuticular wax, the interface between a plant and its environment. This wax is the first barrier to weathering and microbial attack in all plants, and is significant in thatchers' assessment of reed quality (defined as 'sheen' or 'feel'). Unfortunately, relatively little is known about its composition or the factors influencing its development. Waxes are mixtures of extremely stable lipids that vary in composition and quantity depending on age and environmental factors such as temperature, frost, drought and salt concentrations.

No decay can occur on a thatched roof unless there is moisture present for fungal growth. The flow of air and moisture through a thatched roof has been examined by the Building Research Establishment under the guidance of the late Peter Brockett.

Almost nothing is known about the genetic structure of reed populations. It is clear, however, that reed quality can vary drastically from year to year, and within one bed, depending on growing conditions. Some beds never produce poor quality reed, and others never produce good reed. It is quite possible that distinct ecotypes, or even subspecies, of reed exist which differ in characteristics that are significant from the point of view of the thatching industry. At the present time, it is safe to say that reed from around the country, growing in both brackish and fresh water, is of the same species and that differences in stem quality arise due to local climatic and growing/management conditions, although significant genetic differences undoubtedly exist.

ECOLOGICAL ISSUES

Reed beds are ideal habitats for some of Britain's rarest birds, such as the bittern, the focus of much recent conservation work by the Royal Society for the Protection of Birds (RSPB 1994). Bittern nest in lowland marsh dominated by water reed and with ample open pools and old standing reed. They feed at the edge of naturally encroaching reed, and cut reed beds, where water levels are high in the spring and feed primarily on young eels. Managed reed beds with good fish populations are ideal habitats, as long as some standing reed is available for nesting. Reed beds form particularly good habitat if they are cut in small patches, and the RSPB is moving towards increasing its output of thatching reed from managed beds as a positive spin-off from conservation practices to protect bitterns.

There is some concern among naturalists that former peat workings in the Somerset Levels will be replaced by managed reed beds, rather than the natural wetland of mixed vegetation which would otherwise develop. Sedge peat is derived primarily from sawgrass, now extinct in the area, and is richer in nutrients and less acidic than moss peat which makes it more valuable to the horticultural industry. Large reed beds up to 80 ha are being created with grants from the European Community and various conservation bodies. With wetland becoming such a rare habitat in the United Kingdom and most large expanses of fen already drained, there is an argument for maintaining a mosaic of habitats, including managed reed beds, but dominated by wild fen vegetation with the rare species it contains. Reed beds can be created on just about any land, including coarse pasture or low quality arable currently under set-aside, and the return on reed production is almost as remunerative as wheat production.

OTHER AQUATIC MATERIALS USED FOR THATCHING

Sedge used in the thatching industry is more precisely saw sedge (*Cladium mariscus*), a stout perennial member of the *Cyperaceae* family that can grow to 3 m or more in good conditions (Clapham *et al* 1987, 593). It is particularly plentiful in the Norfolk Broads, and is rare and scattered elsewhere due to the destruction of wetland habitat. Sedge is evergreen and spreads like reed by a creeping rhizome and forms dense mono-dominant stands, particularly on fine organic, neutral or alkaline soil in still, shallow water. It has long and extremely sharp serrated leaves, which makes it impossible to apply without thick leather gloves. Sedge is maintained as a complementary crop to reed, often adjacent to it in beds. Unlike reed, sedge may be cut at any time of the year, but it is best cut in late July, which kills encroaching vegetation and lets more growth occur before winter. Commercial sedge is cut every four years in order to produce sedge of sufficient length for thatched ridges. If cut too often, other plants will encroach and the sedge will be of insufficient length. In contrast to reed beds, sedge beds are drained in summer for cutting, and some beds in the Fens are known to have been cut regularly for centuries.

The common club or bull rush (*Schoenoplectus lacustris*) is no longer used for thatching, but was widely used in the medieval period. Like reed and sawgrass, it too is a rhizomatous perennial that grows to 3 m in shallow water up to 2 m deep, in sluggish streams and rivers, and in ponds and lakes particularly with silted bottoms. It is very common in such habitats in south-eastern and central England, but is scattered and local elsewhere.

The black bog rush (*Schoenus nigricans*), a densely tufted perennial with tough wiry stems, rarely exceeds 750 mm in height and was formerly maintained in beds judging by samples obtained from nineteenth century buildings in Leicestershire and Cambridgeshire. It is more exacting in its habitat requirements, and prefers damp, peaty, alkaline locations near the seas and occasionally in salt marsh. It too is widely distributed, but rarely abundant, throughout Britain.

3 Thatching straw

John Letts

BOTANY

Straw is a superbly strong structure for its weight and density and a marvel of natural architecture (Juniper 1990). Essentially it is composed of a cylinder of thick-walled (ie lignified) sclerenchyma cells that provide most of its strength, interspersed with vascular bundles (water- and nutrient-conducting vessels). Some of the parenchyma that forms the soft tissue (pith) in the centre of the stem can also be lignified to some degree, but in general the pith, whether solid, semi-solid or absent, has little impact on the strength of the mature straw. Stems are stronger towards their base, and weakest at the nodes where the strengthening is provided by lignified collenchyma cells.

The epidermis (or skin) is made up of a relatively thin layer of photosynthetic tissue with an external coating of epicuticular wax. It is obviously the first barrier to moisture penetration and disease and pest attack, and remains the first barrier to moisture penetration when the straw is used for thatch. The wax itself is composed of very small semi-crystalline particles made up primarily of long-chain alkane, ester and alcohol waxes and may also have anti-fungal properties. Little work has been done on the composition of these waxes, or on the factors that affect their development in cereals. In general, winter wheats are waxier than spring wheats, and older varieties are often waxier than modern ones.

The epidermis also contains variable amounts of hard silicon crystals embedded in its tissues. Although these crystals may play a role in limiting pest damage, they do not seem to play a major structural role and are not a critical factor determining the quality of thatching straw.

DECAY

A thatched roof is composed of an exposed outer layer subject to weathering, a middle layer with active invertebrate and fungal activity, and an inner zone too dry for fungal activity. Research at the University of Bath (Kirby & Rayner 1989) has attempted to unravel the dynamics of the fungal community of this middle layer, in order to understand the factors that triggered the accelerated decay that appeared on some combed wheat reed roofs in the 1970s and early 1980s.

Fungi are the most important microbes involved in the decay of thatched roofs. The fungal community in thatch is composed of both ascomycete and basidiomycete fungi similar to those found in terrestrial litter systems. Premature decay, first reported officially in 1965, was probably the product of rapid fungal growth of a fungi named *Haplographium*, which degrades cellulose and is common in decaying bark.

Fungal communities can develop in many directions, and fungi adopt different life strategies depending on the resources that are present. Their growth is influenced by variables such as temperature, pH and the type and amount of nitrogen present, but the rate of fungal decay depends primarily on the presence of moisture, which is dictated by thatching techniques as much as any other factor. The faster a roof dries out after being wetted the slower it will decay. Any factor that slows the drying process will affect the rate of decay. Nitrogen levels are also important in determining the structure and activity of the fungal community. High nitrogen levels encourage the growth of fungi such as *Haplographium* that are more efficient degraders of straw.

Tests on decay which were to examine the effect of thatching methods on decay as well as substrate characteristics were initiated at the University of Bath in the 1980s, but were not completed due to a lack of funding. The preliminary results of this study were unable to link straw characteristics such strength and flexibility to variations in decay rates in the laboratory.

THE ASSESSMENT OF STRAW QUALITY

Thatchers have clear opinions as to what constitutes high-quality thatching straw, but their assessments are subjective and difficult to relate to characters that are measurable in the laboratory (Kirby *et al* 1990). They involve visual and tactile assessments of simple physical attributes such as coarseness, hollowness, straightness, length and colour, as well as complex mechanical factors such as strength, flexibility and hardness. Durability is always the most important character defining good quality thatching reed, however, and is also the most difficult character to assess for it is conditioned by a complex array of technical, environmental and biotic factors which pertain to the finished roof as well the characteristics of the reed itself (All references to 'reed' in this chapter refer to wheat, rather than water reed, a traditional thatching usage).

For combed wheat reed, thatchers want tall, straight, evenly tapered reed free of leaf and weeds. The ideal reed stands between 1–1.2 m in the field, producing a cut length of 0.9–1.1 m. Reed is generally considered unusable if it is less than 700 mm long, and most of the reed used falls between 750–900 mm. Short reed is thought to be less durable (perhaps as a result of having been cut higher on the stem), and packs flatter than taller reed when applied to the same depth, thus reducing the pitch of the material and encouraging water infiltration.

Most thatchers prefer hollow-stemmed reed and claim that pithy straw is less durable because it absorbs water and thus decays more quickly in wet weather. This opinion seems to be most entrenched among thatchers who used the semi-solid variety *Capelle-Desprez* as combed wheat reed or long straw in the 1950s and 1960s, for very few thatchers alive today have used true solid-stemmed wheats. *Capelle*'s poor performance was not due simply to its pithy stem, and pithy wheats may not perform very well as long straw or on tightly-packed, shallow-pitched roof in the West Country, but they may have performed adequately as apex-flailed, butts-together long straw on steeply pitched roofs in the Midlands and East Anglia. In all cases, however, the nodes must be solid and without discolouration or decay.

Pale coloured straw with a dull shine and a waxy feel is usually preferred, for glossy reed is thought to indicate the use of fungicides and nitrogen fertiliser. Older wheats tend to be waxier than newer varieties, but this varies with growing conditions. Herbicides and fungicides can decrease wax production, whereas plants will secrete more wax to protect themselves during cold winters, drought or when growing conditions are not ideal. Waxy straw will shed water more readily on the roof and absorb less water when being yealmed or wetted prior to being applied as long straw or combed wheat reed. Waxiness is probably an important, but often overlooked, factor contributing to longevity, particularly in the West Country where roofs can remain wet for several months at a time.

Most thatchers agree that coarse combed wheat reed is more durable than fine reed, although the latter produces a better finish. Strength is thought to reflect durability, and is measured by twisting a few strands into a coarse rope and pulling until it breaks. Brittle reed will produce a frayed rope that breaks easily, whereas durable reed produces a strong and flexible rope. Hardness is also thought to reflect durability, and is assessed by the stem's resistance to crushing (especially at its butt). Good combed wheat reed also has a distinct smell, without the mustiness that indicates storage in damp conditions or the sweet smell of straw that has not properly matured in the barn. Good quality long straw shares many of the characteristics of good combed wheat reed, but perhaps with a greater emphasis on length, flexibility, colour and workability.

PLANT BREEDING BEFORE 1940

Pure varieties of domestic wheat, largely derived from old English land races, dominated wheat production in the main cereal-growing districts by the late nineteenth century, while mixtures of such varieties predominated in areas less committed to intensive cereal production. Many of these were selections from tall, hardy land races of *Squarehead* wheat, a group of old English bread wheats (*Triticum aestivum*) with stout, 1.2–1.5 m straw and dense ears, that produced respectable yields of soft grain well-suited to traditional bread and biscuit production. The *Squarehead* variety most widely cultivated in 1900 was developed from a single plant discovered in a field of *Victoria* wheat in 1868, and is the English grandparent of most of the wheats of significance to thatchers and grain producers today (Percival 1943, 109).

A second species of wheat commonly known as rivet, pollard or cone wheat (*Triticum turgidum*) was also grown until the late nineteenth century (op cit, 89). The rivet wheats were generally taller, higher yielding and more resistant to lodging than the bread wheats, particularly when grown on heavy soils in southern England. Their straw was thick-walled with solid or semi-solid pith, and although much in demand for thatching was unpalatable to livestock and too stiff to be used as bedding or litter. Rivet wheats were also very resistant to disease and pest attack, but many were sensitive to frost and all were very late to ripen. Rivet flour was soft, fine-grained, rich in bran and ideal for use in pastries and biscuits, but lacked the gluten required for commercial bread production.

Hardy land races of bread and rivet wheat were widely cultivated in the early nineteenth century, but were rarely encountered by 1900. With hindsight not available to scientific improvers of the day, their disappearance was an immense loss to plant breeding for they contained the distillation of centuries of crop evolution. In such genetically diverse crops, plants with morphological and physiological characteristics that allow them to set seed in all but the most disastrous of seasons contribute more offspring to succeeding generations. High-yielding characteristics may be present, but it is their association with other adaptive characters such as disease resistance, cold tolerance or strong stems that allows a yield advantage to be expressed. Semi-natural selection within land races generally encouraged the multiplication of taller plants that were able to trap more sunlight, and thus mature more seed, than short plants. But taller plants are also more susceptible to lodging in high winds and wet weather, particularly when grown in fertile soils which encourage rapid stem elongation and produce heavier ears. As a result, the grain output of a land race could never be raised simply by increasing soil fertility. Over many generations land races evolved that were genetically heterogeneous, well-adapted to local growing conditions, variable in height and ripening time and which produced reliably low yields of grain on straw that could reach 2 m in average conditions. Such long straw was welcomed by thatchers and on farms where straw was required for bedding and fodder, but it was an obstacle limiting significant advances in grain output prior to the introduction of semi-dwarf varieties in the 1950s.

In the late nineteenth century pure varieties of wheat were developed by private breeders in response to the lucrative and rapidly expanding urban markets for bread. The rapid advances in yield and grain quality initially achieved were almost inevitable given the genetic variability of the existing crop, and the increasingly scientific methods of selection employed by breeders such as Le Couteur and Shireff. In general, however, progress in plant breeding was hampered by primitive field and laboratory methods, and by poor understanding of the laws of inheritance.

The shift to mechanization after 1850 provided a new, and much more powerful, incentive for breeders to develop varieties better adapted to mechanical reaping and threshing. Mechanization demanded the use of pure crops of even height that ripened evenly, threshed efficiently, and had stiff straw resistant to lodging. The development of reliably shorter varieties gradually became as important a goal for plant breeders as improvements in grain quality and yield. Most of the reapers and threshing machines used in Britain from the late nineteenth century until the introduction of the combine were physically unable to process crops much taller than 1.4 m or shorter than 0.6 m.

IMPROVEMENTS IN GRAIN QUALITY

State-funded plant breeding stations were established within university departments at Cambridge, Aberystwyth, Edinburgh and Belfast before 1900. Roland Biffen, appointed to the Department of Agriculture at Cambridge in 1896, applied the new knowledge to English wheat and developed the pedigree method of selection that proved so successful in succeeding decades (Palmer 1970, Bingham et al 1991).

Biffen's main task was to improve the baking quality of English wheat. In 1900 English varieties were the highest yielding in the world and produced flour with good flavour, texture and colour. Their flour was also low in gluten, however, and produced dense, unmarketable loaves compared to bread made from high-gluten, hard-grained wheats imported from Canada and other regions with hot, dry summers. Improving the quality of British wheat was essential for the survival of the wheat growing industry. A Home Grown Wheat Committee was set up in 1901 by the National Association of British and Irish Millers, and worked closely with the university-based plant breeding stations and the Board of Agriculture to assist breeders in developing high-quality baking wheats adapted to English conditions. The demand for fine white bread made from imported grain was so great that by 1921 home-grown wheat was unmarketable in most parts of England, and growers were unable to sell their finest English grain to millers even in time of shortage (Halliwell 1905).

Research had already demonstrated that grain quality in wheat was primarily an inherited characteristic rather than environmentally determined (Biffen 1926). In the early years of the new century, Dutch breeders crossed a *Squarehead* selection with *Talavera*, an old English variety similar to *Squarehead*, to produce *Victor* (1908), a high yielding, white-grained wheat with stout 1–1.1m straw that was recommended by the National Institute of Agricultural Botany for general cultivation until 1957 (and which is still occasionally grown for thatching). Two other selections from the *Squarehead* group were of particular significance to later breeding programmes: *Squareheads Master* (synonymous with *Red Standard*) was selected from *Squarehead* by a private breeder in the West Midlands in the 1880s, and *Browick*, which Biffen crossed with *Red Fife* in 1916 to produce *Yeoman*, the first high-yielding, high-quality baking wheat available to English growers (Lupton 1987).

Squareheads Master inherited its hardiness and tall 1.1m straw from its English ancestor, and yielded reliably well in a wide range of soil and climatic conditions. It was the most popular winter wheat in Britain until the late 1930s, and was recommended by the National Institute of Agricultural Botany for cultivation in special circumstances until 1960. Unfortunately, its grain was soft and produced an inferior loaf, and the pedigree of much the seed sold by growers and seed merchants was also suspect. Biffen crossed *Squareheads Master* with the Russian wheat *Ghirka* in 1910 to produce *Little Joss*, a higher yielding variety with slightly taller straw and good disease resistance. *Yeoman* inherited its grain qualities from *Red Fife*, the Polish spring wheat that had turned the Canadian prairies into the breadbasket of Europe, and its stout straw from *Browick* and *Squarehead*. *Yeoman* set a new quality standard for English baking wheat and remained on the National Institute of Agricultural Botany Recommended List until 1957.

THE ORIGIN OF PLANT BREEDERS' RIGHTS LEGISLATION

Records indicate that an average Victorian farm might grow ten or more named varieties in any given year, in order to match varietal characteristics to local growing conditions. Bratton Farm in Wiltshire grew at least 24 different named varieties between 1840 and 1850 (Morrison 1993) and 85 varieties of significance were listed in the *Cyclopedia* of 1856 (Morton 1856). This diversity persisted into the 1930s when Percival listed 80 varieties of bread and rivet wheat still in cultivation in Britain, although many on a restricted scale. Much of this diversity was genuine, and stands in stark contrast to the varietal uniformity of the present day, but many improved varieties were marketed for the benefit of the seed dealer and were identical to varieties already grown under another name. Rigorous trials comparing the performance of mixed varieties with confused names were of little use to growers, particularly since the seed of most varieties was unavailable on a national basis.

In response to this confusion, many growers and breeders lobbied for the adoption of a registration system for pure races of wheat as existed in several other European countries. A registration system already existed in the United Kingdom for pedigree farm animals, and it was argued that a similar system for plant varieties was essential for future progress. Many breeders and growers were opposed to such a system and argued that the pedigreed varieties available were so similar to the mixed races already grown that the legislation was unnecessary. Some actively opposed the spread of pedigree varieties on sound agronomic principles, arguing that common mixtures used the resources of a field more completely and performed better than the improved varieties in most situations.

At the outset of the First World War the bulk of English wheat seed was home-produced, usually by farmers on contract to seed merchants. It varied in quality with the pedigrees of many popular varieties clouded by synonyms. The outbreak of the war prompted the passing of the Testing of Seeds Order (1917), which was consolidated in the Seeds Act (1920) and the Seeds Regulations (1922) which tried to bring order to the chaos that was seen to exist in the seed trade (Wellington 1966). A further decade passed before any more steps were taken in the direction of Plant Breeders' Rights legislation. In the meantime, stringent legislation to protect the unsuspecting grower was being passed abroad, such as the Seeds Act (1923) in Canada (Engledow 1923). This legislation made it illegal to misrepresent the character or quality of a new plant variety, and prevented its marketing until it had been tested for 'purity and distinctiveness' and licensed by the Minister of Agriculture. Similar registration programmes were gradually adopted by agricultural societies and governments on the Continent, and the fledgling United Kingdom seed industry lobbied to bring in a registration system in the United Kingdom.

The first British national seed certification scheme (for white clover) was sponsored by the Ministry of Agriculture in 1930, and in 1932 a Cereal Synonym Committee appointed by the Royal Agricultural Society, the National Institute of Agricultural Botany, the National Farmer's Union, and the National Association of Corn and Agricultural Merchants published the first list of cereal synonyms (Bell 1934). The confusion was particularly intense within wheat; *Squareheads Master*, *Wilhelmina* and other very popular varieties each had two or more widely cultivated forms that were almost identical. New varieties were clearly a gamble, and a grower could change seed several times without result. The National Institute of Agricultural Botany's Farmer's Leaflet No 1 published in 1930 provided growers with their first reliable objective assessment of the varieties available on a national basis. The leaflet argued that by growing the 'correct' variety for local conditions a farmer could expect a 20% increase in yield. The programme was immediately successful in encouraging the use of recommended varieties, and Leaflet No 1 gradually evolved into the National Institute of Agricultural Botany Recommended List of Cereal Varieties.

DEVELOPMENTS DURING AND AFTER THE SECOND WORLD WAR

At the outbreak of the Second World War, most farms still depended on manuring and crop rotation to maintain soil fertility, but artificial fertilisers were already widely used. Lawes' early research on phosphate and nitrogen supplements at Rothampstead had led to the use of various natural minerals and manufacturing by-products as fertilisers, and to the creation of an artificial fertiliser industry in the early 1900s (anon 1931). Increases in yield due to intensive fertilisation inevitably led to lodging, and selection for shorter stems within the European wheats usually reduced plant vigour and yields at the same time. Plant breeding had absorbed many lessons from the emerging field of genetics, however, and the link between height and yield was soon to be broken. Improved hybridization techniques had even producing inter-species crosses such as *Triticale* (*Triticum* x *Secale*), the wheat x rye cross that is a major cereal crop and a significant source of thatching straw. Advances in crop protection had also kept pace, but the thatcher's desire for quality straw was rarely the subject of attention.

The bulk of the wheat crop was produced from a dozen or so relatively pure varieties, rather than the plethora of mixtures and impure selections cultivated at the start of the First World War, and included:

1) *Squarehead* (mid 1800s)
2) *Squareheads Master* (1880s) – syn *Red Standard*
3) *Red Marvel* (1904) – syn *Red Japhet*
4) *Bersee* (1935)
5) *Starling* (1907)
6) *Victor* (1908)
7) *Wilhelmina* (1910)
8) *Little Joss* (1910)
9) *Yeoman* (1916)
10) *Vilmorins' Hybrid 27* (1927)
11) *Rampton Rivet* (1934)
12) *Juliana* (1936)
13) *Holdfast* (1936)

Only one rivet wheat, *Rampton Rivet*, a medium-height, hollow-stemmed selection made by Frank Engledow at Cambridge, survived in cultivation on heavy clay soils in southern England.

The demand for high quality certified seed of pedigreed varieties continued into the post-war period, and breeders redoubled their efforts to improve the yield and quality of British wheat in a post-war era of shortage and rationing. The United Kingdom cereal seed market was soon swamped by foreign varieties (Gill & Vear 1969). *Bersee* (1935) was very popular, but like many of the new varieties performed poorly in poor soils and low input systems. *Bersee* was used as breeding stock for many of the wheats that dominated cereal production in the United Kingdom in the 1950s and 1960s such as *Hybrid 46* (1953), *Capelle-Desprez* (1953) and *Elite Lepeuple* (1960). *Hybrid 46* had short, stiff straw and was particularly popular in the fertile soils of the Fens in the 1960s. *Capelle*'s popularity also peaked in the mid 1960s, at one time covering 80% of United Kingdom wheat fields. It remained on the National Institute of Agricultural Botany Recommended List for 24 years, and ten of the 15 varieties on the Recommended List in 1976 were derived from it. *Capelle-Desprez* was crossed with *Holdfast* to produce *Maris Widgeon* (1964), and with *Hybrid 46* (1946), *Professeur Marchal* (1960) and a Japanese dwarf wheat *Norin 10* to produce *Maris Huntsman* (1972) and *Kinsman* (1976) in the early 1970s. *Maris Huntsman* performed well in low input systems and poor soils, but its straw was considered to be tall, thin-walled and susceptible to lodging.

Genes for dwarfing were first discovered in Japanese spring wheats in the 1940s, and were incorporated into European wheats in the 1950s and 1960s. Dwarfing genes allow plants to funnel photosynthate to the ear rather than the stem, and in highly fertile conditions maximize yields by limiting stem elongation while increasing the number of grains in the ear. The first semi-dwarf varieties were hailed as miracle grains and were marketed aggressively in the 1960s and 1970s throughout the world by advocates of the green revolution. Dwarf varieties, however, do not perform well in poor soils and in low input systems, and the green revolution was more successful in eroding genetic diversity and destroying traditional agricultural systems in both developed and developing countries than in feeding the hungry. By the 1980s, 80% of winter wheat acreage in Britain was planted to semi-dwarf wheats, and they have dominated the National Institute of Agricultural Botany Recommended List ever since.

Scientific plant breeding had relatively little impact on the thatching industry prior to the Second World War, but as breeders focused on reducing plant height as the key to raising yields, thatchers increasingly found their previous supplies of straw too short for combing or too damaged by the combine for use as long straw. A demand developed for specialist straw producers, and as threshing machines were abandoned on cereal farms, they were taken over by small farmers who were unable (or unwilling) to compete with large grain producers, and by thatchers who expanded into straw production in order to guarantee their access to raw material.

THE ADOPTION OF PLANT BREEDERS' RIGHTS AND THE NATIONAL LIST

The movement that eventually led to the adoption of Plant Breeders' Rights legislation in Britain can be traced to the activities of the International Association of Breeders for the Protection of New Plant Varieties. After its foundation in 1901 as the public voice of the plant breeding industry, the International Association of Breeders for the Protection of New Plant Varieties helped establish voluntary varietal regulation systems in Germany, France and Holland and several other European countries. The issue of Plant Breeders' Rights in the United Kingdom was brought up at a conference in Stockholm in 1954, in response to a survey sponsored by the Organisation for European Economic Co-operation on seed production, testing and distribution (Howard 1960). The survey report concluded that crop varieties were as patentable as any other invention, and recommended that member countries introduce a system of registration for new varieties, and act to ensure that plant breeders obtained a proper reward for their efforts.

The United Kingdom Committee on Transactions in Seeds presented its general report in July 1957. The Committee recommended that the government enact legislation to encourage the development of a private plant breeding sector in friendly rivalry with state-funded institutions. This would depend on the granting of proprietary rights to the breeders of new varieties, so that other breeders and growers would not be able to sell seed of a licensed variety without the permission of the owner. A breeder would have to demonstrate that a new variety was 'distinct, uniform and stable' before it could be registered, after which time royalties could be collected on the sale of seed for a period of 15 years. New names would have to conform to the rules of the International Code of Nomenclature for Cultivated Plants, and varieties would be protected in signatory states by an International Convention for the Protection of New Varieties of Plants. Discussions eventually led to the Plant Varieties and Seeds Act (1964), and to the creation of the various organisations to administer the Plant Breeders' Rights programme which came into operation in 1965 (Howard 1967). An Index of Names of Plant Varieties was also started in April 1966, which lists the names of all licensed varieties grown in the United Kingdom, and ensures that new varieties are sold under only one name.

A breeder is given an exclusive legal right to collect royalties from the marketing of a new variety once a grant of Plant Breeders' Rights has been issued, but the actual licensing for the purpose of collecting royalties is voluntary. In contrast, the National List, introduced in 1974 after Britain's entry to the European Economic Community (now European Union), is compulsory, and has had an immense impact on agriculture and the thatching industry. The legislation requires member states to maintain a National List of varieties whose seed can be marketed legally within the Community. If the variety does not appear on the list, its seed cannot legally be sold, traded, given away or in any way have its ownership transferred for the purposes of cultivation. Varieties are not listed until they have been shown to be 'distinct, uniform and stable', and to possess 'value for cultivation and use'. Systematic evaluation of new varieties is sponsored by the Ministry of Agriculture, Fisheries and Food and undertaken by the National Institute of Agricultural Botany in harmony with existing testing for grants of Plant Breeders' Rights and for inclusion on the National Institute of Agricultural Botany Recommended List. The new legislation was initially designed to ensure that EEC farmers were supplied with authentic and reliable seed as a means of improving grain production within the community, in the days before grain mountains and set-aside. It prevents the marketing of varieties with 'limited' economic potential or serious agronomic defects, and a stated aim is to 'give breeders and the seed trade the widest opportunity to commercialise new varieties within the community'.

In the first two years of National List testing, new varieties are compared for 'value for cultivation and use' against older established varieties in order to assess their relative merit in terms of basic agronomic characters and market acceptability. 'Value for cultivation and use' is less difficult to establish than distinctiveness, uniformity and stability, which examines morphological, genetic and physiological characters in detail. The descriptions produced at the end of the second year are registered by the Seed Certification Authority and for the grant of Plant Breeders' Rights. To be accepted onto the National List, a new variety must be judged better than the lower half of the varieties already on the list, so that only half of the new varieties submitted are ever accepted onto the list. Varieties are listed for ten years, and if the breeder does not renew its registration by paying a fee the variety is dropped from the list (MAFF, pers comm). The National List contains 80 varieties, and the 1994 version included both *Maris Huntsman* and *Maris Widgeon*, two of the most popular varieties used in thatching.

THE NATIONAL INSTITUTE OF AGRICULTURAL BOTANY RECOMMENDED LIST

The best varieties from the National List enter the National Institute of Agricultural Botany Recommended List trials, which are funded by levies collected from growers and the seed trade and are administered by the Home Grown Cereals Authority. These tests are also undertaken at the National Institute of Agricultural Botany, but with much more stringent selection criteria for factors such as disease resistance and grain quality. Successful varieties are given Provisional Recommendation after two years of trials, and eventually General Recommendation at the end of an additional year of testing under all of the growing conditions to which they could be exposed in the United Kingdom. To be accepted onto the Recommended List a new variety must be as good, or better, than the best on the current list, and in practice only one in ten ever receives full recommendation.

In recent years, varieties have remained on the Recommended List for approximately seven years, and over 90% of the United Kingdom cereal acreage is planted with recommended varieties. In spite of the higher initial purchase cost of their seed, recommended varieties are rapidly adopted by growers and older varieties gradually become outclassed and fall into obscurity. In the 1994 Recommended List, six varieties were given general recommendation, six were provisionally recommended subject to further testing and four previously recommended varieties are soon to be outclassed by new claimants. The cost of developing, testing and licensing a new variety prohibits individuals or small organisations from taking on the task, and any new tall-stemmed 'thatch' varieties are unlikely to pass the National Institute of Agricultural Botany's tests or be granted Plant Breeders' Rights. Even though practical experience and historical data suggests that high quality thatching straw might best be obtained from a land race, or mixtures of numerous older varieties (or even species) grown in conditions that are totally unsuitable for modern varieties, it would be illegal to market its seed or give it away to a grower dedicated to improving the quality or output of their reed crop.

The National and Recommended Lists are widely seen by grain producers and breeders as essential to the continued vitality and competitiveness of United Kingdom agriculture, and the profits generated by Plant Breeder's Rights legislation have encouraged the development of a vigorous private plant breeding industry.

Half of the massive increase in yield that has occurred since the Second World War is thought to be due to the introduction of new varieties, particularly the modern semi-dwarfs with stiff straw that are able to convert massive applications of artificial fertilisers into grain in ideal conditions. Nitrogen inputs increased from 47 to 80kg/ha between 1947 and 1965, to 100kg/ha in 1975 and almost 200kg/ha in 1989. Although necessary to maintain current levels of yield and quality, additional inputs are not expected to contribute significantly to future improvements. Breeding effort is being directed to improving varietal resistance to biological and climatic hazards, and to 'tailoring' varieties to local growing conditions in order to improve the reliability of yields. Twentieth-century plant breeding has all but exhausted its conventional gene pool of advantageous characteristics, and breeders are increasingly turning to wild relatives of domesticated crops, and to biotechnology, for useful genes that can be incorporated into existing varieties rapidly and profitably.

VARIETIES USED FOR THATCH

There is considerable confusion among legislators, thatchers and the public as to the origin, botanical characteristics and qualities of the many wheat varieties available for thatching. Press articles often describe long straw and combed wheat reed as wheat varieties, and few thatchers can name more than three or four varieties that produce good thatching straw. The significance of varietal change has also been missed by scientists involved in thatch research. A survey conducted in the late 1980s concluded that 'varietal differences do not appear to be important factors in thatchers' assessments of reed quality'. It is clear from the unpublished results, however, that the comparison was unreliable as it was made with samples of only three closely related varieties (*Maris Huntsman*, *Maris Widgeon*, and *Squareheads Master*) of unrecorded provenance. All growers and most thatchers agree that the choice of variety grown is crucial, but unfortunately no reliable data is available comparing the qualities of straw from different varieties grown in controlled conditions.

With the exception of *Squareheads Master*, the old English varieties had largely been supplanted by the hybridised products of scientific plant breeding such as *Little Joss* (1910), *Victor* (1908), *Wilhelmina* (1910) and *Yeoman* (1916) by the late 1920s. This first generation of improved hybrids produced relatively good quality grain and excellent thatching straw, but were in turn supplanted by shorter, stiff-strawed varieties such as *Bersee* (1935), *Holdfast* (1936) and *Juliana* (1936) which produced higher yields when grown with artificial fertiliser, and could be harvested more efficiently by combine. Older varieties persisted in areas less committed to intensive cereal production, particularly in the West Country and on smaller farms in areas where thatching remained a popular roofing material. The older varieties produced better quality combed wheat reed, but many of the newer varieties produced acceptable long straw when threshed. Varietal selection may have been less of a concern in eastern England where a thatched roof could be expected to last longer simply due to drier conditions.

The varietal schism between straw growers and grain producers widened in the decades after the Second World War, and has continued to widen into the present day. *N59* (1959), *Elite Lepeuple* (1960) and *Chalk* (1962) produced good quality combed wheat reed and are still grown on a small scale. Others including *Capelle-Desprez* (1953), *Hybrid 46* (1953) and *Professeur Marchal* (1960) were very popular with grain producers, but were never widely accepted as thatching wheats. Derivatives of *Capelle-Desprez* fared much better: *Maris Widgeon* (1964), *Maris Huntsman* (1972), *Aquila* (1976) and *Kinsman* (1976) rose to prominence as thatching wheats in the early 1970s, due in part to the premium which could be obtained from the sale of their grain for high quality feed. Although they perform well in low input systems and on poor soils, they were soon outclassed for intensive grain production. Others, such as *Maris Dove* (1971), *Maris Freeman* (1974), and *Flinnor* (1974) were briefly popular but are rarely grown.

A limited survey undertaken in 1977 (Pullen 1979) revealed that over half of the straw used for combed wheat reed was produced from *Maris Huntsman* and *Maris Widgeon*, the former being the most widely grown winter wheat in England until the early 1980s. *Capelle-Desprez*, *Chalk*, *Flanders*, *Bouquet* and *Flinnor* made up a further 38% of reed production, with the remaining 9% derived from *Maris Ranger*, *Elite Lepeuple*, *Professeur Marchal*, *Maris Dove* and *N59*. When grown without nitrogen fertiliser *Maris Huntsman* rarely exceeds 0.9 m, which is close to the minimum height required for thatching. All of the wheats developed since the 1970s have even shorter straw than *Huntsman* and are not suitable for thatching.

Maris Huntsman and *Maris Widgeon* yield approximately 1.5–2 tons/acre of combed wheat reed (*c* 80–120 28 lb nitches/acre or 3.8-5 tonnes/ha) or long straw, and an equivalent amount of grain, whereas older varieties produce much less grain but higher yields of straw.

RYE AND *TRITICALE*

Rye (*Secale cereale*) is a minor crop in Britain, with production concentrated in relatively poor, sandy soils in Norfolk, Suffolk and the Vale of York. It was once widely cultivated in England, due to its hardiness, ability to grow in acidic conditions, the quality of its straw and its nutritious grain. English rye will grow to 1.8 m, taller than almost any bread wheat. Its straw is similar to winter wheat, but thicker and more flexible. Traditionally, winter rye was sown in early autumn and was commonly interplanted with early ripening wheat, whereas spring-sown rye is a mid nineteenth-century introduction. Rye was very commonly used for both combed wheat reed and long straw style thatching, and is still used for ridging in various parts of the country, but it is generally felt to be softer and less durable than wheat straw.

Triticale, a cross between durum wheat and rye, inherited the latter's ability to grow on poor, acidic soils and in colder climates. It has an aggressive root system and tall, stiff straw that rarely lodges. Some thatchers claim that it is the most durable thatching straw available, whereas others view it with suspicion on account of its hybrid origin.

GROWING THATCHING STRAW IN ENGLAND

In the past growers of combed wheat reed were obviously aware that certain varieties or mixtures of wheat produced better quality reed. In general the portion of the crop destined for use as thatch differed only in the greater care that was taken not to damage its stems during harvest, and the fact that it was combed rather than threshed. Similarly, most of the wheat varieties grown prior to the Second World War produced acceptable long straw, but the quality of the straw available depended on the extent to which it was damaged and mixed during threshing.

As long as the wheat crop was cut with a binder and threshed by hand or by machine, straw suitable for long straw-style thatching was abundant and inexpensive, particularly in the main cereal growing districts such as East Anglia, the Midland plains and the chalk downs of Wiltshire, Berkshire and Hampshire. In general, machine-threshed straw was crushed along its entire length whereas hand-flailed straw was crushed only where the flail struck its blow. But in the past as in the present day, threshing machines varied in the degree to which they damaged the straw, as did the skill of the crew employed to run it.

The quality and availability of long straw remained steady through the late nineteenth century and into the early decades of the twentieth century, but crisis struck in the late 1930s and 1940s as larger cereal producers intensified production and adopted combine harvesters and shorter, brittle-stemmed varieties better suited to modern farming.

As combine harvesters spread through the countryside, threshing machines were abandoned except for those in the hands of a dwindling number of farmers who were unable, or unwilling, to compete with large grain producers, but who could produce high quality combed wheat reed or long straw for thatching. Many thatchers also acquired threshers (and combers) to be guaranteed access to adequate material at a reasonable price. In the early 1950s the Rural Industries Bureau reported that the spread of the combine was seriously undermining the quality and availability of combed wheat reed and long straw (Rural Industries Bureau 1952, 1954–55). Prices and demand were steady, but shortages often occurred due to poor growing conditions, in part a reflection of the limited acreage planted with the older varieties. But as the thresher increasingly became a tool for thatchers to remove grain from thatching straw, rather than simply a mechanized flail that happened to produce a by-product that could be cleaned and used for thatching, the quality and quantity of long straw increased. This is supported by Rural Industries Bureau records, and by the responses to a questionnaire sent to thatchers in the Cambridgeshire region in early 1994 (Stanford 1994). By the early 1960s the widespread shift to combine harvesting had severed the link between the farmer and the thatcher, who relied on the contractor and dealer for most of his straw (Rural Industries Bureau 1961–2). Since few combers existed outside the West Country, the export of combed wheat reed to other regions where the Bureau had been encouraging its use led to rising prices and shortages of long straw.

Since the 1960s, combed wheat reed production has become concentrated in the hands of fewer, larger, specialist producers in the West Country, and production has increased in many other areas of southern England. An upswing in the demand for thatched roofs has been countered by poor publicity arising from 'premature decay', which for some has destroyed confidence in the durability of straw thatch and increased the demand for water reed. At the same time, rising labour costs, and the perennial shortage of labour at harvest time, has pushed straw producers to the limit of efficiency, and few growers are expanding production even though good quality combed wheat reed commands a fair price in the marketplace.

The premium received from the sale of threshed or combed grain is more significant than many admit, but a dedicated grower's main goal is always the production of high quality straw regardless of grain yield. The reed producer's greatest fear is lodging, but other problems such as dog-leg (bent basal internodes resulting from steady spring winds and/or soil compaction), goose-neck (ears bending over when mature), 'take-all' and eyespot (and other fungal diseases that cause discolouration and lodging) can greatly reduce the straw's market value and the grower's reputation. The best reed is produced on fairly heavy, well-drained soils with low to moderate fertility. Poor soils often produce quality crops, and growers agree that 'some fields always produce good reed, while others never will'. In summary, quality control depends very much on knowledge and experience. As in the past, new producers entering the market for quick profits are notoriously unsuccessful, for in contrast to modern dwarf wheats older varieties vary tremendously in character, growing requirements and response to inputs.

The best reed is derived from slow-growing winter wheat, harvested before it is fully ripe in late summer and ripened for several months in the stook and barn before being threshed or combed. Before the emergence of specialist straw producers in the 1950s and 1960s, straw from older spring wheats such as *April Bearded* was used when winter straw was unavailable, and could outperform winter wheat when well-applied. It is no doubt the case that the best spring wheats produce better thatching straw than the worst winter wheats, particularly when grown without artificial fertilisers.

Before the widespread use of artificial inputs, soil fertility, weeds, pests and diseases were controlled by judicious crop rotation, tillage, fallow, hand weeding. A second crop of wheat in the same field usually failed, as occurred during the Second World War when government advisors eager to increase cereal production imposed wheat quotas on farmers and ignored existing rotations. Modern intensive cereal production demands maximum yields from the same plot of land year after year, and success depends on high inputs of artificial fertilisers and chemical sprays rather than 'natural' methods of control.

In southern England thatching wheat is usually sown in late September or early October, and seeding rates depend on the tillering ability of the variety sown. Dense planting will encourage competition for light and lead to lodging, whereas most varieties will tiller profusely to make up for thin planting.

Few issues attract as much controversy among growers and thatchers as artificial fertilisers, and particularly the impact of nitrogen on reed strength and durability. Nitrogen is essential to plant growth, and high levels will stimulate photosynthesis and increase yields of both straw and grain, but the resulting straw may be susceptible to premature decay when used for thatch. In the past, high fertility was rarely a problem for wheat producers except in naturally fertile soils such as the Fens. Fields were not manured for wheat production, and wheat usually followed a root crop or a clover ley (as is still the custom amongst reed growers in Devon). Autumn cereals were planted early and their well-established root systems mopped up the soluble nitrogen that would otherwise be leached from the soil by winter rains. Precocious growth in the autumn or early spring was grazed by sheep or scythed to reduce the risk of lodging and to promote evenness and tillering.

Lodging has become a rare problem for wheat producers since the introduction of stiff-strawed dwarf wheats in the 1960s, and the large increases in grain yield in the 1970s and 1980s were achieved through massive applications of nitrogen fertilisers. Even minor applications of nitrogen will stimulate growth and produce a taller, heavier and, if the crop does not lodge, a more profitable crop. Most growers add 1.5–3 cwt per acre (188–376 kg per hectare) of a low nitrogen (25:50:50) compound fertiliser at seeding time. Many find additional top-dressings of nitrogen difficult to resist and add 50–75 kg per hectare of nitrate in the early spring depending on conditions (MAFF recommended the higher value in the 1970s). The highest quality reed is said to be grown without any nitrogen supplements after seeding, but unfortunately for the buyer, reed grown with moderate (but undesirable) amounts of nitrogen looks and feels very similar to reed grown organically. No reliable tests have been done to compare the relative qualities of organic straw with straw grown with moderate levels of nitrogen. The University of Bath research suggested that nitrogen was concentrated in the butts of straw grown with artificial fertilisers but accumulated in the ears and upper internodes of straw grown organically (Kirby & Rayner 1989). Few stems were tested, however, and the samples were mainly of unknown provenance. The tests also ignored differences attributable to storage, timing of harvest and variety. Nitrogen does appear to reduce the thickness of the sclerenchyma ring that gives straw its strength, even though stem diameter may not be affected.

Some concern was expressed in the 1970s about the impact of herbicides and fungicide sprays on the quality of thatching reed, but again the evidence for reductions in quality is anecdotal. Herbicides are used routinely by growers of combed wheat reed and long straw, and many also apply fungicides at least once during the growing season. Research indicates that herbicides reduce the production of epicuticular wax, the plant's first barrier to moisture penetration, fungal attack and other environmental hazards. Fungicides are often considered essential to combat the host of fungal diseases that can destroy a reed crop overnight, but research indicates that fungicide sprays also destroy beneficial root mycorrhiza (symbiotic fungi) that help plants extract nutrients from poor soils, and to deal with environmental stresses such as drought or parasitic fungi (Barea *et al* 1993). Along with residual nitrate in straw, fungicides may alter the ecological balance in the fungal community normally present on a thatched roof, and indirectly contribute to accelerated decay.

Thatching reed is cut with binder leaving a 100 mm stubble. The timing of the harvest is crucial, and a crop can be at its optimum harvest quality for only 24 hours. Combed wheat reed is harvested several weeks before what would be considered optimal for harvesting by combine, for thatching straw must be flexible and lignification proceeds rapidly in the last few weeks of growth. The ears must be erect and the upper nodes still green, with a rainbow of colours separating the golden stem from the ripening ears. Tests for ripeness often include drawing a thumbnail up the stem to detect moisture. The grain must be yellow ripe or cheesy, whereas grain producers combine their crops when dead ripe. According to many of the 'old boys' who farmed in 1930s and 1940s, younger farmers are afraid to cut early and let the grain ripen in storage, even though the grain will ripen perfectly well in the stook. Sheaves are stooked for two or three weeks and combed directly from the field if labour is available, or are stored in a barn or rick until sufficient labour is available. Traditionally, the best reed was kept in the rick until Christmas, or until it had lost its sweet harvest smell.

CONCLUSION

The last five decades of agricultural 'progress' and government legislation have not benefited the growers of thatching straw. The priority since the First World War has been the creation of high-yielding, disease-resistant, dwarf wheats responsive to artificial inputs, and straw producers have either had to maintain their stocks of older varieties or else use the best of a bad lot of newer varieties with poorer quality straw.

Although there is a steady and significant demand for quality thatching straw, neither the Ministry of Agriculture, Fisheries and Food (MAFF), the National Institute of Agricultural Botany nor the Council for Small Industries in Rural Areas have ever undertaken a systematic trial of thatching wheats. MAFF's lack of support may be due to the fact that such trials would obviously have to include non-listed wheat varieties, which they would not be able to encourage due to United Kingdom and EEC rules governing the sale and distribution of non-listed varieties. MAFF's trial conducted in Devon in the 1970s was inconclusive because the trial roof was rethatched just as differences in varietal performance were becoming obvious, and the varieties tested were no longer commercially available (or even on the National List) by the time the test was discontinued (Pullen 1979). MAFF continues to view straw as a by-product of the grain industry, and the fact that no published data on thatching straw is available is surprising considering that industry generates sales of several million pounds per year (Staniforth 1979).

Plant Breeders' Rights legislation has not been a big problem to thatch straw growers, for it mainly affects dwarf varieties that are of no use to thatchers in any case. National listing, however, has discouraged the use of older wheat varieties and has forced reed producers into an EEC-designed corner. The National List prevents the 'transfer of ownership' of seed of any unlisted variety. It is therefore illegal for a grower to give seed of an unlisted variety to a neighbour who wishes to grow reed, or whose crop has been entirely destroyed by disease or bad weather. It is technically illegal for researchers to transfer ownership of seed obtained from gene banks to growers who wish to experiment with older varieties in order to improve the quality of their straw. The legislation actively works against the maintenance, and expansion, of a high-quality thatching straw industry.

The legislation also prevents impure, or mixed, varieties from being licensed. In the past, high quality reed was obtained from genetically mixed crops that ripened unevenly, whereas modern thatching wheats are genetically pure and ripen very uniformly. As a result, a labour shortage at harvest time can be disastrous for reed quality, and synchronous ripening is a major bottleneck preventing any expansion of production. A crop composed of mixtures of pure varieties selected for reed quality would ripen unevenly, and might reduce the labour crisis by spreading the harvest period over several weeks. Such mixtures would fail the National List's three main rules: that licensed varieties be distinct, pure and stable. Designed and accidental mixtures are already widely grown for thatching. In the 1970s, varietal mixtures were mooted as a means of reducing the susceptibility of cereal crops to fungal diseases, but similar licensing difficulties were encountered. In the early 1980s some MAFF researchers also argued for the retention of tall, older varieties of wheat on the Recommended List for specialist purposes, even though they had clearly been outclassed by higher yielding dwarf wheats. Older varieties perform better in low input systems and on poor soils, and would have balanced the List for grain producers and encouraged the retention of older varieties on the National List. Many of these older varieties might produce more durable straw than the handful of varieties currently grown.

The preliminary results of three years of growing trials of several hundred older wheat varieties undertaken by the present author at the University of Reading identify three dozen varieties that appear to be well suited to thatching, and field trials of fifty leading lines are already underway in the West Country. Although many researchers at the National Institute of Agricultural Botany and MAFF are sympathetic to the needs of the thatching industry, the legislation appears inflexible as it has widespread support within the agribusiness community.

A proper assessment of the comparative qualities of old English spring wheats, ryes and *Triticale* would also be of great interest to thatchers, straw growers and the owners of thatched properties. Research in many scientific fields such as genetics, mycology, biochemistry and taxonomy could benefit the thatching industry, but the results of such research rarely trickles down in a way that is of interest or use to thatchers. For example, research in Canada has demonstrated that older varieties of bread wheat are better able to form beneficial associations with a range of symbiotic root fungi (mycorrhiza) that improve nutrient uptake in poor soils and low input (ie organic) systems, and also buffer the plant when it is under stress from disease or drought (Hetrick & Wilson 1992). Most modern varieties, and particularly the high-yielding dwarf wheats, have lost this ability and their success depends on high inputs of artificial fertilisers, fungicides and pesticides. Selection for mycorrhizal response could theoretically lead to the identification, and improvement, of older varieties that are able to produce greater yields of high-quality grain without the use of artificial fertilisers. An interdisciplinary study bringing together experts and research from various fields would undoubtedly succeed in improving the durability of thatching straw, and thereby increase the longevity of both combed wheat reed and long straw roofs.

Part II
Organisation and training

4 The Rural Industries Bureau and its successors 1940–94

The Second World War was a major break in the continuity of the thatching trade. Post-war demographic and social changes affected the building industry (with which house-thatching was reluctant to be classed) as a whole, but house-thatching was peculiarly sensitive to both and was also profoundly affected, as it still is, by additional factors: changes in farming practice and the management of wetlands producing water reed. In the period 1940–94 house-thatching was also drawn into a world of business and bureaucracy that was unknown before the war. A series of government organisations, starting with the Rural Industries Bureau, took an interest in thatching and helped to develop the trade that exists today.

THE PURPOSE OF THE RURAL INDUSTRIES BUREAU

The Rural Industries Bureau was founded in November 1921. The Act of Parliament that gave birth to it, the unlikely-sounding Development and Road Improvements Funds Act of 1909, stressed the value of assisting and developing agriculture and 'rural industries'. For reasons of the wording and interpretation of the Act, this was implemented, over a period of time, as assistance directed towards the smaller rural industries.

From 1940 the Bureau was maintained by a grant from the Development Fund, made by the Treasury on the recommendation of the Development Commission. It consisted of a body of Trustees, a Council and administrative and technical staff. In the first year of the war, in September 1939, the Bureau had only twenty members of staff, including clerical assistants. Activity after the war increased and by 1948 there was a staff of forty. By 1954 there were one hundred and thirteen (Williams 1958, 1–17). The Bureau's headquarters was based at Taunton until 1944, when it moved to Camp Road, Wimbledon Common. Thatching was only one of the Bureau's interests from 1940. It was also concerned with (among others) the underwood industry, basket-making, the Welsh woollen industry, saddlery and (small-scale) agricultural engineering.

W M Williams, looking back on the Bureau's activities in 1958 in a book associated with the radical experiments in rural regeneration at Dartington, Devon, argued that most of its governing principles to that date were contradictory. The assumption behind the establishment of the Bureau was that rural areas needed revitalising to counter a demographic shift to urban areas; 'the drift from the country'. The Bureau was dedicated to modernization and progress. This was not applied, however, to a radical view of the future of the countryside, but remained shackled to a romantic myth about rural England, and the assumption that properly rural industries were small and mostly craft-based. This was an outlook that elevated the small-scale rural craftsman in particular to a position of social and economic importance that he had probably only ever occupied in imagination. Sir Basil Mayhew, writing the Bureau's Annual report of 1955 as Chairman of the Trustees, praised the rural craftsman.

> In a world in which local materials and characteristics, all-round skill and individual needs tend to be submerged, they represent qualities, which many nations beside our own have worked hard to preserve, knowing that without them industry, however prosperous, could give a drab uniformity to town and countryside. (Rural Industries Bureau 1955, 4)

Small was not only beautiful, but morally superior.

> Only the very best of the small individual producers – those rather solitary craftsmen who place quality above profit and personal gain, should be supported on the score that they are the setters of standards. The larger producers (and by those I mean employers of perhaps ten or more men) are apt to lose a grip on standards because of the hurly-burly of conducting a business over which they have but little personal supervision. (Williams 1958, 12)

The Bureau's commitment to modernization and progress was shared by many who, after the upheaval of the war, were looking for a better economic and social order. What constituted 'progressive' from the Bureau's perspective however, was not necessarily 'progressive' to all. Viewed from some quarters, such remnants of the vernacular craft and building traditions that did survive after the war and with which the Bureau became involved, were no more than reminders of an unprogressive, old-fashioned world.

THATCHING AND THE BUREAU

Thatching must have been a particularly difficult task for the Bureau to tackle. On the modernization side, there was virtually nothing that could be mechanized once the thatching material was on site and the thatcher at work. Undertaking research required finding thatchers at a time

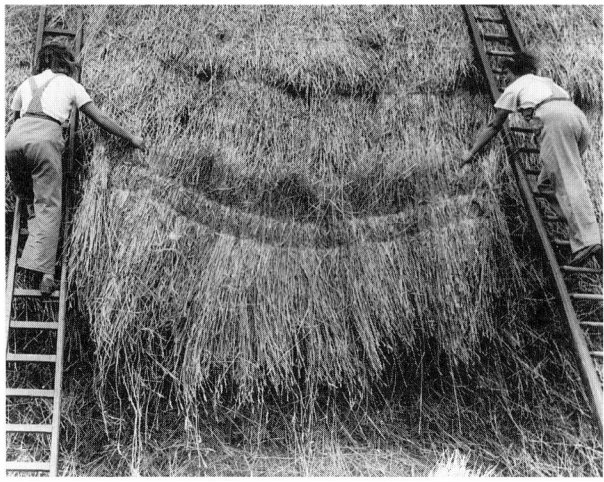

Figure 4 Land girls using machine-stitched thatch for ricks, 1940s (Associated Press).

when telephones were relatively rare and advertising their skills was done largely by word of mouth. Having found them, interviews in the cause of better understanding sometimes had to be conducted up a ladder, and the tendency to put down competitors, little different among craftsmen of all trades in the 1930s and 1940s from today, had to be sifted out of the information. Eliciting information about how thatchers priced work and their profit margins was no easy task. In retrospect, the persistence of the Bureau seems remarkable. The criticism made by Williams (1958, 168) that the statistics and surveys it produced were inaccurate, is important but, under the circumstances, inaccuracy was understandable.

In 1937–8 a survey (Public Record Office 4, 60–62) for the Bureau indicated that there were about 517 house thatchers using straw. Wages were comparable to, but slightly better than those of agricultural labourers. One third of the work force consisted of men over 60, with only 10 apprentices and seven other workers below the age of 30. In addition to straw thatchers, twenty-three firms using water reed were recorded. There was already a recognised problem in the quality of straw for thatching. There was no trade body. There was wild inconsistency in pricing, ranging from a low level from which it appeared impossible to derive a living wage, to prices that the market could not stand. Oral history records evidence and some continued use of a range of locally available thatching materials: rye, heather, sedge, linseed and others. The Bureau (and the organisations which replaced it) played a significant role in a number of changes between the 1937–8 survey and 1994.

Thatchers in the early 1990s were still small in number. Rough estimates suggest approximately 800 thatchers, thatching being worth approximately £45,000,000 per annum. The largest firm had 22 (franchisee) thatchers operating in England, the smallest consisted of one-man bands. There was still a problem with the availability and quality of straw. There were two trade bodies. Prices varied from region to region but a successful thatcher could expect to make a living which produced an income far superior to an agricultural labourer. There were two common thatching materials, wheat straw and water reed.

THE IMPACT OF THE SECOND WORLD WAR

During the war many thatchers found themselves classified as farm workers and key workers in the production of food. This was a sore point for some who, given the choice, would have preferred fighting service to farming service (Jeff King, pers comm). Those that remained at home found themselves occupied with agricultural thatching on the additional acreage of land turned over to grain for feeding the nation. Land girls were drafted in to fill the gap created and assisted with agricultural thatching (Fig 4). The maintenance of roof, rather than rick, thatching was confined to farm buildings where agricultural produce was stored, leaving the maintenance of thatched farmhouses and cottages on hold. In some cases property owners kept these waterproof with galvanized iron sheets, sometimes fixed over existing thatch, leaving, along with those thatched roofs that survived with minimum maintenance, a huge backlog of potential house thatching in the immediate post-war period.

Before 1945 the Bureau had recognised the likely impact of the war on the various rural trades and resolved to take action.

> Craftsmen in the countryside have endured six hard years and the return to their former occupation by many of the younger men serving in the Forces was doubtful. It was obviously of great importance to see what could be done to bring fresh blood into the various trades. (Rural Industries Bureau 1939, 22)

In 1945 both the Oxford and Cambridge Rural Community Councils approached the Bureau, complaining of an acute shortage of thatchers in those areas (Public Record Office 5). Oxford had complained of the same problem in the 1930s and this had produced the 1937–8 survey of thatchers for the Bureau, which this research has failed to trace, although some of the results do exist (Public Record Office 4). The pre-war survey, set against the urgent post-war requests for help, gave the Bureau the impetus for action. Here was potential rural, craft-based employment with little evidence that it could look after itself, but no shortage of available work.

The Bureau appointed its first thatching instructor (the terms 'instructor' and 'officer' became interchangeable in the Bureau's records), Fred Davies, on 21 January 1946 as part of a policy to remedy the shortfall in the number of tradesmen. Like the other technical staff appointed by the Bureau (and this made admirable good sense), Mr Davies was a practising craftsman: 'rather than a school-trained instructor' (Rural Industries Bureau 1939–47, 25), although it must be said that no school-trained thatching instructor existed in 1946.

In March 1946, Cosmo Clark, the Bureau's administrative manager, reported that Davies had been to Oxford to work with the local Community Council and the local Rural Industries Organiser to survey house thatchers in Oxfordshire, Buckinghamshire and Berkshire. These had been found (no easy task), and graded according to their abilities, 'good', 'medium', and 'poor'. Those that were willing to take apprentices were noted. It was suggested in 1946 that county thatching officers should be appointed to supervise training schemes. However, when Oxford's War Agricultural Executive Committee applied for permission to appoint a county thatching officer, this was refused on the grounds that it was not the function of the War Agricultural Executive Committee to train rural craftsmen (Public Record Office 5). If the scheme for county thatching officers had not foundered on a technicality, regional variation and local distinctiveness in thatching might have survived better until the 1970s and 1980s, when they began to be consciously prized.

Failing county thatching officers, the Bureau pressed on with another form of available training scheme and, acting as the Ministry of Labour's agents, proposed to introduce a Vocational Training Scheme to be supervised by its own thatching instructor. As with other trades this was for eligible young men, particularly ex-Service men. Before the scheme could begin, the Bureau came up against the absence of a thatchers' trade organisation of any kind. This snag meant that the Bureau itself had to prove to the Ministry of Labour that training was needed: 'information which would normally be provided by a trade organisation' (Rural Industries Bureau 1947, 24). The Bureau decided to put the trade organisation problem to rights by setting about forming a Master Thatchers' Association. As it turned out, this ambition was not fulfilled until the 1960s; until then the trade organisations were regionally based.

REDEFINING THATCHING

The Bureau's approach to the Vocational Training Schemes was very much of its time. Nearly fifty years later, the stated aims read as a period piece, a mixture of paternalism and faith in progress, with a dash of social theory.

> It is concerned with the recruitment and instruction of new workers, the instruction of practising craftsmen in their own crafts and the conversion from old ideas to new. In providing a technical service for the craftsman the Bureau inevitably becomes involved in the social problem of raising his status. The two tasks are complementary. Any attempt to separate the technical aspect from the social is both impractical and undesirable. (Rural Industries Bureau 1947, 25)

The 'new ideas' and change in status as applied to thatchers, particularly to straw thatchers, were radical.

Most straw thatchers, as the pre- and post-war surveys showed, had earned their living before the Second World War by mixing thatching with other employment. Most of these were seasonal agricultural occupations; hurdle-making, sheep-shearing, hedging and ditching, although car hire was another variation more in tune with progressive thinking. Even most water reed thatchers, who were seen as different altogether in 1946, mixed their business with basketry or making osier hurdles. The Bureau, however, made a clear distinction between agricultural and house thatching:

> It should be stressed that the thatchers with whom the Bureau is concerned are the roof thatchers – not the farm hands who can carry out rick thatching ... Contrary to general belief, a thatcher's trade is not seasonal: he can be fully occupied the whole year round. (Clark 1947, 444)

Looking back, nearly 50 years later, one might question the Bureau's identification of house thatching in 1946–7 as an independent industry at all. The division made by the Bureau between the rick thatcher, who, as they themselves pointed out, was 'found everywhere' and house thatcher, is a curious one at first sight. Some thatchers before the war did indeed thatch only ricks and clamps on which a rough and temporary thatch was adequate for a season. A rick thatcher might not thatch barns or houses and some barn thatchers did not thatch houses. There was a locally accepted understanding of levels of competence and experience (Jack Dodson, pers comm). Nevertheless, any straw thatcher who thatched houses in the 1930s was almost certainly also in the business of agricultural thatching: covering barns or other farm buildings from time to time, or covering ricks as the seasonal need arose. At first sight it is odd, given the enormous pool of rick thatchers, that the Bureau did not concentrate on drawing these wholly relevant agricultural skills into the house-thatching trade, but the priorities were of their time and the obvious pool of unemployed labour on which to draw was ex-servicemen, rather than rick thatchers.

However, bureaucracy itself forced the Bureau to make a distinction between rick and house-thatching. Seasonal occupations, as it happened, did not qualify for the Ministry of Labour's Vocational Training Scheme, and rick-thatching was undeniably seasonal. Before the Vocational Training Scheme, thatching fulfilled neither of the requirements for introducing such a scheme, requirements that were considered to define an industry. Thatching did not have a trade 'voice' and its history in the period before the Second World War was as part and parcel of seasonal work, with the exception of one or two large firms using Norfolk reed. On the first count, the Bureau bent the rules of the Ministry of Labour, who accepted that the Bureau itself was in a position to request training, in the absence of a trade body. On the second count, the Bureau claimed that house-thatching was an activity separate from seasonal thatching, in order to establish a training scheme (Clark 1948). In short, in 1946–7 the Bureau could be said to have made official a specific house-thatching industry that included thatchers using straw as well as water reed.

Meanwhile, experienced thatchers who were not engaged in the Bureau's training scheme were coping with the post-war changes and thatching in altered circumstances. Jack Dodson had learned long straw thatching before the war from his father, who worked on an estate in Huntingdon, thatching ricks and houses, but in an independent capacity rather than as an estate thatcher. The son returned from a year in the army and set up Dodson Brothers, who continued to work on the estate when called in, but also thatched for private customers.

A van inherited from their father and the post-war shortage of thatchers gave Dodson Brothers a wide market, which they approached in an exceptionally business-like manner, including requiring written contracts, interim payments and long working weeks. The firm travelled further and further afield, including Scotland and Worcester. They stayed in digs when they were far from home, and would arrange next year's work in advance (Jack Dodson, pers comm).

Sid Pearce also learned long straw thatching from his father, who worked on a large Wiltshire estate. Before the war he and his brothers mixed sheep-shearing with thatching ricks, farm buildings and houses. After wartime service he found that his job on the estate had been filled by another thatcher and he and his brother set up in business in Pewsey Vale, mostly thatching houses, not ricks (Sid Pearce, pers comm).

Alan Fooks' family had worked as thatchers on an estate in Dorset and he also learned to thatch in long straw from his father, thatching a mixture of ricks, farm buildings, estate cottages and houses. In 1948 the estate was sold for death duties and the estate thatchers were obliged to go independent, continuing to thatch in the same area, but with new occupants of the buildings, which had been sold to tenants, less able to keep up the standard of their roofs than the estate had been. Thatching in a limited area he was able to continue to use a bicycle and trailer, the straw (which had previously been supplied and delivered by the estate) being delivered to the sites where he worked by an agent (Alan Fooks, pers comm).

The practicalities of finding work after the war for these three successful thatchers, all of whom thatched in straw to begin with and found full-time employment as house thatchers, substantiated the 'official' birth of house thatching recognised as an independent industry by government departments.

One side effect of the separation of house and agricultural thatching, along with increasing costs and the divorce of straw thatchers from working farms, was the drastic loss of thatch from farm buildings. As enterprising craftsmen, with or without the Bureau's help, got to work re-thatching dwellings, the stock of thatch on farm buildings fell into further disrepair. This was a continuation of the loss of thatch on farm buildings, a pattern that dated from the agricultural depression of the 1870s. Yards of previously thatched buildings received corrugated iron or, later, corrugated asbestos coverings or were re-roofed in other materials. This loss is most obvious in old photographs, and has continued, unlikely ever to be reversed except when farm buildings are converted into houses.

Early training and the Bureau's thatching officers

Although training schemes for the other trades with which the Bureau was involved were undertaken at their headquarters at Wimbledon Common, thatchers were taught on site. The apprenticeship scheme in response to the shortage of thatchers in Oxfordshire and Cambridgeshire was supervised by Davies, the Bureau's thatching instructor, soon assisted by others. The training period was 39 weeks, with the employing thatcher being paid 7s 6d (37p) for the first 17 weeks of the trainee's employment, but thereafter gradually contributing to his wages, as he or she became more useful. After training was completed, the master thatcher was required to keep the trainee in work for a year, and pay the current Agricultural Wages Board rate, 80s per week (Public Record Office 6). The only aspect of the system which had to be revised was the selection of trainees. The Ministry of Labour and National Service saw the scheme as an opportunity to mop up unemployed ex-servicemen. As it turned out, sending mature men into the thatching trade did not prove attractive to master thatchers.

> Most Master Thatchers prefer the young boy who has not acquired a set way of life and who comes from the country or even from the district in which he is to be trained. (Rural Industries Bureau 1950, 22)

Women were also recruited to thatching, and one, trained under a disabled persons' scheme, wrote a revealing account of the shortcomings of the training as well as her views on the necessary personal qualities needed by a thatcher (see Appendix 2).

The Bureau's thatching instructors also gave on-site instruction to experienced thatchers, including demonstrations, at their own request (Fig 5). This began what was, overall, a subtle relationship between Fred Davies and his successors at the Bureau, and the industry. The Bureau was the first body that had anything approaching an overview of thatching across the country and their officers were the first individuals who saw the practice and diversity of the craft at work from East Anglia to Cornwall. This was a national perspective that automatically led to a notion of 'good' and 'bad' thatching by comparing areas that had never been considered side by side, and that paid little attention to the historic reasons for the diversity of materials, or the historic reasons for the variable durability of thatched roofs.

The thatching officers deserve a special mention. There have only been nine in all since Fred Davies' appointment in 1946; William Martin, Fred Cooper,

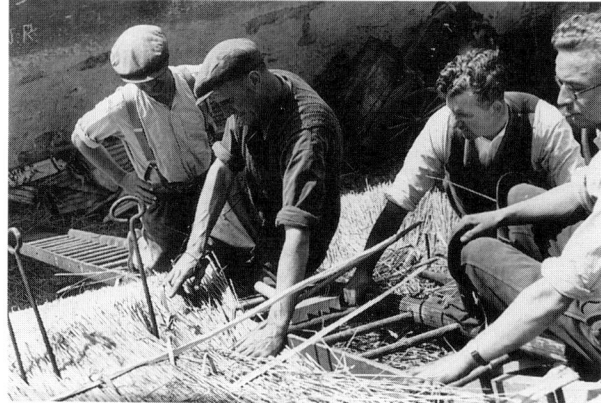

Figure 5 William Martin, the Rural Industries Bureau's South West region thatching officer instructing master thatchers in reed wheat (combed wheat reed) thatching, 1949 (Rural History Centre, University of Reading).

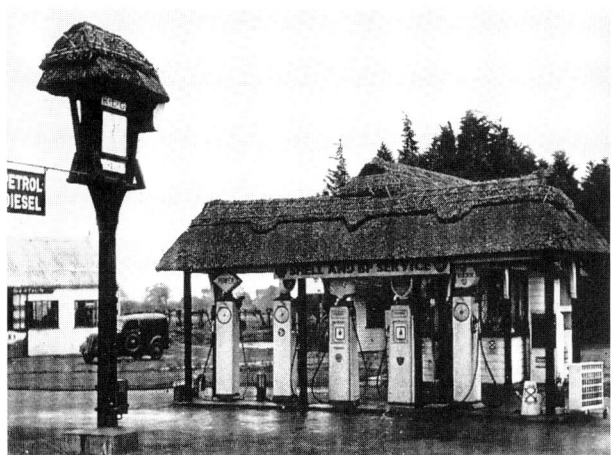

Figure 6 Thatch put to new buildings. A thatched petrol station in Essex, 1950s, thatched by Derek Wisbey (Derek Wisbey).

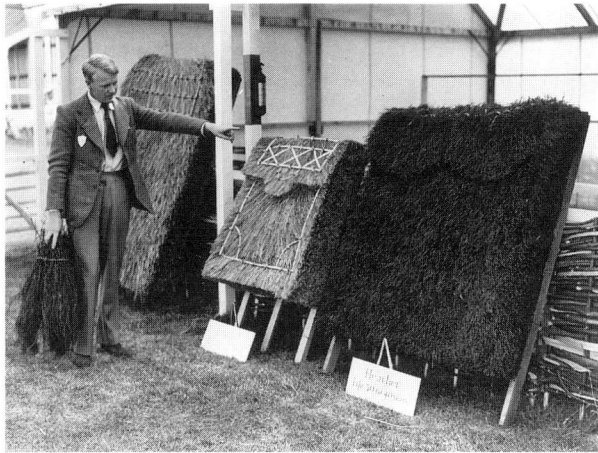

Figure 7 The Rural Industries Bureau display at an agricultural show, 1949 (Farmers' Weekly/Rural History Centre).

George Dalliston, Reg Clarke, Derek Wisbey, Jeff King, Peter Brockett and Paul Norman. The Bureau's officers, and those that later worked for the Bureau's successor organisations, the Council for Small Industries in Rural Areas and the Rural Development Commission, were intermediaries who straddled, sometimes uncomfortably, the divide between bureaucracy and the thatchers themselves. In one respect they were government officials, part of a bureaucratic centralised organisation, preoccupied with paperwork, and different from working thatchers for that reason. They advertised thatch by thatching. Derek Wisbey, in particular, thatched not only a petrol station (Fig 6) but also a soft-topped sports car (in water reed), which became a familiar sight in the Cambridgeshire area. They spoke at conferences (Public Record Office 7), and demonstrated thatching to the public at agricultural shows (Fig 7). They were also different because they had a perspective broader than the next roof on the books.

The thatching officers were bounced into a position, known to no other thatchers, of seeing, and puzzling over, the regional differences and individual quirks of practitioners. This made them men apart, and even sometimes in opposition, as they encountered individuals who were thought not to come up to standard. As 'government officials' they took a great deal of criticism from thatchers, including accusations of favouritism in the days when they were allowed to recommend craftsmen. In other respects they were all practical thatchers themselves with a craftsman's interest in solving practical problems. They became intimately involved with trade organisations as these evolved. Fred Cooper, for instance, one of the best-loved and most diplomatic Bureau officers, fetched up as secretary of the National Society of Master Thatchers' Associations in the 1960s, to be followed in this post by Jeff King, also a thatching officer.

Had the thatching officers been bureaucrats or theoreticians only, the Bureau and its successors might have let go of thatching and thatchers, as the industry developed. The officers, however, were a powerful link between the Bureau as centre and the individuals, disconnected from one another, who made up the nascent thatching industry. They were genuinely the friends of many thatchers, had a commitment to the craft and developed their own ideas about it. Help with book-keeping, obtaining supplies of material, dealing with customers that failed to pay their bills or with domestic difficulties were all part of the job. If some fell away, suffering from what would be recognised now as stress-related illnesses, some dedicated themselves to every aspect of thatching and, inevitably, became involved in the complicated politics that were to develop in the matter of trade organisations.

In the post-war period, through the medium of the thatching officers, the Bureau was in a position to be proactive, to be master of the industry, *as* an industry rather than a collection of isolated practitioners, and orchestrate changes in it. This is not to say that the Bureau reached all thatchers, and it is impossible, from its own records, to know how many thatchers fell completely outside its orbit, never joined one of the local trade associations, never asked for instruction, never had any involvement in the assisted apprenticeship schemes and continued to mix house thatching, as they always had done, with other activities, for so long as a diversity of employment made economic sense and was available. Oral and local documentary history are important tools for amplifying the narrative of post-war thatching revealed in the Bureau's records.

MATERIALS ACCORDING TO THE RURAL INDUSTRIES BUREAU

However richly diverse thatching materials and techniques had been before the Second World War and in spite of oral history's evidence of diversity after the war, the Bureau's annual thatching reports from 1950 onwards relate almost exclusively to straw and water reed as materials and to the three main thatching techniques that are widely recognised in the 1990s: long straw, combed wheat reed and water reed. The thatching officers were aware of alternative materials, of course, but from the outset they concentrated their energies on promoting the most common survivors. The perceptions of the different materials and techniques in the Bureau's annual reports on thatching make fascinating reading and are closely intertwined. The two techniques for straw thatching were the first major preoccupation.

The Rural Industries Bureau's perception of long straw, 1947–68

The 1940s

In 1947 the perception of the Bureau was that long straw was a poor performance material, relative to combed wheat reed. Water reed was not a point of comparison for the Bureau at this date. A 1947 article by Cosmo Clark, the Administrative Director of the Rural Industries Bureau, in *Agriculture*, stated that the condition of thatching on houses and farm buildings in Oxfordshire, Buckinghamshire and Berkshire, where straw thatch was long straw, was 'deplorable'. 'The reason for this is that the straw available in these counties does not last' (Clark 1947, 446). The longest life a thatcher could promise for a long straw roof was fifteen years, the average life was ten years and some roofs lasted only as long as five. The costings Clark gave for roof maintenance over 30 years were based on these figures for durability, and meant that costs of maintenance for long straw were calculated at either £249 or £153 (depending on the type) compared with combed wheat at £45, which meant that long straw, at its worst, was judged then to be over five times more expensive than combed wheat reed over a 30 year period. No figures were supplied for water reed.

Clark identified four different estimated costs of straw thatching, based on information compiled by Fred Davies and the Oxford Rural Community Council. The hours quoted were based on actual experience, the wage rates were those of the Southern Counties Federation of Building Trades Employers, January 1946, Grade A.

The definition of long straw has proved something of a poser in the last ten years or so, and Clark's description has been used in some of the recent debates as defining different types of long straw (he himself used the word 'method'). Something of the context in which Clark was writing may help to explain his account.

At the request of the Minister of Works, a committee had been set up by the Royal Institute of British Architects under the chairmanship of Edward Maufe, to study the architectural use of building materials, resulting in *Post-War Building Studies No. 18* (1946). The study expressed some enthusiasm for thatch on the grounds that it was cheap, local, a good insulating material and 'it saves timber and other materials', meaning that re-thatching was more economical than altering a roof for another material. The Committee concluded that 'an effort should be made to bring renewed life to this essentially local craft. A system of apprenticeship should be organised before thatching becomes a lost art'. Clark's article was designed to show that the Bureau was already covering these recommendations. He gave the Bureau's figures from their seven-county survey begun in 1946. The estimates for the maintenance costs of long straw are shown in Tables 1 and 2.

There are some oddities here. Firstly, the difference between the two estimates described as long straw has nothing, apparently, to do with technique. The costs, and presumably the nature of the labour, are exactly the same. The difference between the two estimates depends on the difference between 'loose threshed wheat straw' as opposed to 'best selected wheat straw', and this, in turn, depends on whether the straw is produced directly to the thatcher from the 'property owner' or whether it is supplied by a 'contractor' (it is not clear whether this is an agent, or the thatcher himself). Looking closely at the first estimate, Clark seems to be suggesting that a farmer supplying his own straw was content to use any waste straw for his own house and thus accepted a less durable thatch than a property-owner who was unable to supply the material himself. If this is the case, it is difficult to understand who was being charged 50s per ton for the material.

Table 1 Estimate for long straw thatching, after Clark 1947.

Estimate 1 Long straw method		£ s d
MATERIALS supplied by property owner	loose threshed wheat straw from own rick at 50s per ton life 5–10 years	
	straw, 3 cwt	0 7 6
	wood, hazel or willow	0 2 6
	sub-total	**0 10 0**
LABOUR	ON GROUND straw wetting twice, drawing, yealming and bundling 4 hours labourer's rate	0 8 0
	wood splitting, sharpening and twisting 250 sprays and some runners 3 hours labourer's rate	0 6 0
	ON ROOF laying, fixing and trimming 5 hours craftsman's rate	0 12 6
	sub-total	**0 26 6**
CHARGES	overheads and profit 20% (of labour only)	0 5 0
	total cost per square	0 41 6
	total cost of building of 20 squares	41 10 0
	total maintenance cost over 30 years (renewed every five years)	249 0 0

Table 2 Estimate for long straw thatching, after Clark 1947.

Estimate 2 Long straw method (unclipped surface)		£ s d
MATERIALS supplied by contractor	best threshed wheat straw, selected for thatching, trussed and delivered to site at £4 10s per ton life 10–15 years	
	straw, 3 cwt	0 13 6
	wood, 1 bundle	0 2 6
	sub-total	**0 16 0**
LABOUR	as in Table 1/Estimate 1	0 26 0
	sub-total	**0 42 6**
CHARGES	overheads and profits, 20%	**0 8 6**
	total cost per square	0 51 0
	total cost of building of 20 squares	51 0 0
	total maintenance cost over 30 years (renewed every 10 years)	153 0 0

Table 3 Estimate for thatching straw laid reed-wise, after Clark 1947.

Estimate 3 Straw laid reed-wise (surface clipped)		£ s d
MATERIALS supplied by contractor	best threshed wheat straw, selected for thatching, trussed, delivered to site at £4 10s per ton life 15–20 years	
	straw, 3 cwt	0 13 6
	wood, 1 bundle long hazel rods	0 2 6
	sub-total	**0 16 0**
LABOUR	on ground as Table 1/ Estimate 1	0 14 0
	on roof laying, fixing, trimming and clipping surface 10 hours craftsman's rate	0 25 0
	sub-total	**0 39 0**
CHARGES	overheads and profits 20%	**0 11 0**
	total cost per square	0 66 0
	total cost per building of 20 squares	66 0 0
	total maintenance cost over 30 years (renewed every 15 years)	132 0 0

Table 4 Estimate for thatching straw laid reed-wise, after Clark 1947.

Estimate 4 Straw laid reed-wise, butt ends on surface and beaten into place (no clipping)		£ s d
MATERIALS supplied by contractor	best unthreshed wheat straw, stripped and trussed, known as 'Devon reed', at £11 per ton life 30–40 years	
	Devon reed 3 cwt	0 33 0
	wood, 1 bundle long hazel rods	0 2 6
	sub-total	**0 35 6**
LABOUR	on ground	0 2 0
	straw tying into small bundles (no wetting or yealming) 1 hour labourer's rate wood splitting, sharpening and twisting 250 sprays and some runners 3 hours labourer's rate	0 6 0
	on roof laying, fixing and levelling 7.5 hours craftsman's rate	0 18 9
	sub-total	**0 26 9**
CHARGES	overheads and profit 20%	0 12 3
	total cost per square	0 74 6
	total cost of building of 20 squares	74 10 0

If the thatcher were paid a small maintenance fee and travelling expenses he would visit the work twice a year and plug any weak spots with 'bolts' of Devon reed as required. The total maintenance cost over 30 years should not exceed £45.

What was Clark driving at in these compared estimates? The crucial difference seems to be the expectation and demands of an owner who could supply his own straw, and one who could not. Farmers and large landowners had had a more or less continuous involvement with thatchers carrying out agricultural as well as house-thatching. They may have been content with material of relatively short duration on their houses, because it was practical to use up waste straw and no great inconvenience to carry out patching or re-thatching on the house or cottages, simply as and when it was needed, using labour to hand.

An owner without the advantage of his own straw to hand was in a different economic context. Here, the thatcher had to make a special visit to carry out the work, and a job of a different quality would have been required. It is also worth remembering that Clark, who was writing in answer to an Royal Institute of British Architects committee, must have been keen to establish that good quality thatch could be produced in a context where modern architects could consider choosing it and specifying for it in comparison with alternative roof coverings. Durability and cost came to the forefront. Durability had not always been the prime consideration when straw thatch was a farm-based material and both straw and thatchers were to hand on the farm. The cost of house-thatching in straw was likely to have been lost in wages bills to employees who sometimes thatched houses, but sometimes carried out other tasks for the farmer.

Clark's third estimate (Table 3) is worth considering here, for reasons that should become plain. This estimate is clearly neither long straw, in Clark's definition of the first two estimates, nor combed wheat reed, although it is described as straw laid 'reed-wise'. The straw has been threshed and yealmed, the surface not 'beaten into place' (see Table 4), but trimmed and clipped. Apart from the clipping (which is done by some, but not all, long straw thatchers in the 1990s), the technique of thatching described here would seem to correspond roughly to 1990s long straw thatching, not to any 'reed' type of technique, as it would be understood today. The figures given for longevity are far better than the 'long straw' described in the first two estimates. The material and method of acquiring it is perceived to be the same as in Table 1, the only difference in described technique is the addition of clipping the surface.

Fred Davies and the Rural Community Council for Oxford, who provided Clark with the information, may have been doing no more than describing something (and probably not all) of the diverse methods in using threshed straw he encountered, which are rather confusingly mixed up in the estimates with the perception of differences between clients, whether farmers or non-farmers. Clark himself, no thatcher, may simply have been muddled by the information with which he was supplied.

Intentionally or not, the ambiguities in the description of the thatches above pinpoints one of the most interesting aspects of thatching, that is the extent to which the actual production of materials, both physical and in terms of relationships, contributes to the identity of the three common thatches. It could be argued that the thatching style of the straw thatches begins in the field, according to variety, how it is harvested and then processed and transferred from person to person, from field to roof.

The Bureau identified straw produced for long straw as a different 'material' from straw produced for combed wheat reed, although there were occasions when the varieties of wheat were the same. Different varieties used for combed wheat reed compared to long straw thatching were as likely to reflect local availability as to reflect substantially different requirements for the two types of thatch. Letts' discussion of the most suitable varieties for thatching today, in Chapter 3, is written at a time when the survival of suitable varieties is more pressing than it was to the Bureau, who tended to make a simple distinction between long-stemmed varieties they saw falling out of use, and the more 'modern' short-stemmed varieties that were replacing them.

The 1950s
By 1949–50 the Bureau was predicting the disappearance of long straw. Combed wheat was 'likely to replace the long straw method which although cheaper is not nearly so durable' (Rural Industries Bureau 1950, 22). Two years later in 1953, when the Bureau had rather more experience under its belt, the lifespan of long straw had mysteriously risen to 'approximately 20 years', corresponding more closely to Table 3. The Bureau was also suggesting a longer lifespan for combed wheat reed, which was then considered to last from 35 to 50 years (Rural Industries Bureau 1953, 17).

In 1954–5, the Bureau's annual report included reflections on the reasons for having encouraged a change to combed wheat reed from long straw in the previous years. These make very interesting reading and were rather different from those offered by Cosmo Clark seven years earlier. It was not just the poor quality of material that gave some long straw such a short life span, but apparently the poor standards of some long straw thatchers were a problem. The Bureau had realised that in many areas which had been strongholds of long straw thatching:

> individual standards left much to be desired, and that the livelihood of sound craftsmen was liable to be seriously jeopardised by less able men ... it was seen that the introduction of a new technique would enable the instructors to influence standards of workmanship from the start. (Rural Industries Bureau 1955, 20)

It is clear that one of the reasons why the Bureau preferred a fresh start with combed wheat reed was the opportunity for re-educating poor quality long straw thatchers. Doubts about the quality of long straw thatchers in the 1950s may have been the result of diversity of practice in this technique, including thin coats that may not have lasted, or have been expected to last, for very long. Oral history records that, for example, some Hampshire thatchers put on a new coat of thatch that was 5 inches (125 mm) thick, in contrast to others, who might use a coat 15 inches (375 mm) thick (Peter Brockett, pers comm). Thatchers in the early 1990s recollected that their fathers put less material on than their sons do today (Steve Cleeve, pers comm). If particular conventions in the depth of new coats in some areas were associated with scarcity of suitable straw, it would be no surprise that thatching lasted less well. Long straw was the type of straw thatch found in those areas (among others) where cereal production was intensive. These were most responsive to the developments in wheat breeding, to farming dedicated to increased grain yield and to new harvesting machinery. The quality of long straw thatch in these areas was bound to have suffered as a result.

The question of aesthetics was also a factor in the Bureau's preference for combed wheat reed. Long straw 'in the opinion of many' was the least satisfactory in appearance of all the thatches.

> Applied as a flat coating to the roof, it lacks the rich texture created by the tightly-packed butt ends visible when other methods are used. (Rural Industries Bureau 1957, 26)

In 1958–9 it seemed that rye straw might help to fill the wheat straw gap. The Bureau noted that more rye straw was being grown and was suitable both for long straw and combed wheat reed (Public Record Office 8).

The 1960s
The thinking of the Bureau, as shown in the annual reports, became more sophisticated in the 1960s, with a greater understanding of the problems of supply and demand and occasional references to the changing character of the owners of thatched houses.

In 1959–60 there is the first reference to what the Bureau clearly saw as hybrid types of thatch. The Midlands instructor reported that combed straw 'was being used in the Midlands in the long straw manner by thatchers not yet versed in the reed-laying technique'. The promotion of combed wheat reed as a material had clearly overtaken the Bureau's ability to train thatchers in the 'Reeding technique' that they saw as conventionally accompanying straw processed in a particular way. 'This unorthodox practice must bring in train a still heavier demand for instruction in an area where the combed wheat reed technique has been progressively introduced by the thatching officer over a period of years' (Rural Industries Bureau 1959, 34). In the same year came the first reference to sparring water reed onto straw. The straw technique that was being covered over was not identified.

> During the September quarter, however, the thatching officers spent several days with a Welsh thatcher who over the years had experimented with and perfected a method of half-coating with reed on a straw base. In this method the impoverished top layers were stripped from the roof till sound straw was revealed and on to this either Norfolk or marsh reed was secured and the thatch finished to a normal thickness. (Rural Industries Bureau 1960, 35)

Oral history, as in many instances, enriches the Bureau's understanding of the chronology of techniques. At least three generations of the Smith family in Lymington, Hampshire, have spar-coated water reed (Jeff King, pers comm) and so the Welsh example drawn to the attention of the Bureau was not the first.

In 1959–60 the Bureau carried out a survey to identify the numbers of thatched buildings in different counties, their condition and, in some, to identify the relationship between the different types of thatch (Public Record

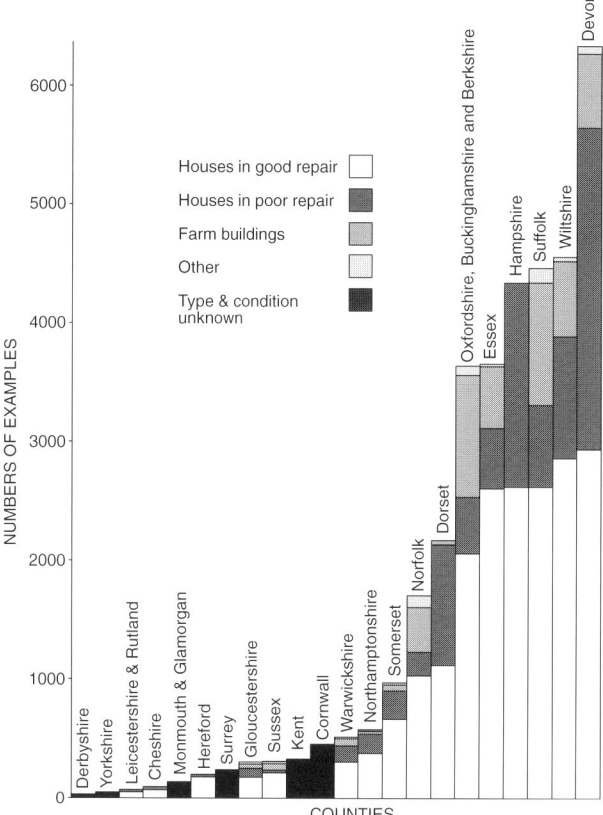

Figure 8 1959–60 Rural Industries Bureau survey expressed as a chart (see Table 5).

Office 17) (Table 5 and Figure 8). It is not clear how or by whom the survey was undertaken. In Devon it was based upon a 15% sample. In some counties the returns were incomplete and some counties were investigated *en bloc*, eg Oxfordshire, Buckinghamshire and Berkshire. Not all the county returns included information on the types of thatch. Nevertheless, the results provide a rough and ready guide to the relative quantity of thatched buildings surviving in different counties, and, in some counties, the relationship between long straw, combed wheat reed and water reed. In the south, which in Bureau terms comprised Hampshire, Sussex, Kent and Surrey, the types of thatch were not identified, neither were they in Yorkshire and Derbyshire which, in Bureau terms, fell into the Midlands and North regions. In the other counties in this region, long straw predominated, in Oxfordshire, Buckinghamshire, Berkshire, Northamptonshire, Warwickshire, Gloucestershire, Herefordshire, Cheshire and Leicester and Rutland. In East Anglia, only Norfolk could claim a dominance of water reed; Suffolk and Essex were both predominantly long straw. In the south west there was a little long straw in Dorset and less in Somerset, but, along with Devon, where it was ubiquitous, combed wheat reed dominated. The figures for Cornwall and Wiltshire were not broken down into different thatching types (Figs 9, 10, 11 and 12). The survey was not mentioned in the annual thatching reports. These became longer, the conclusions less clearcut, especially when shortages occurred all round, as they did in 1961–2.

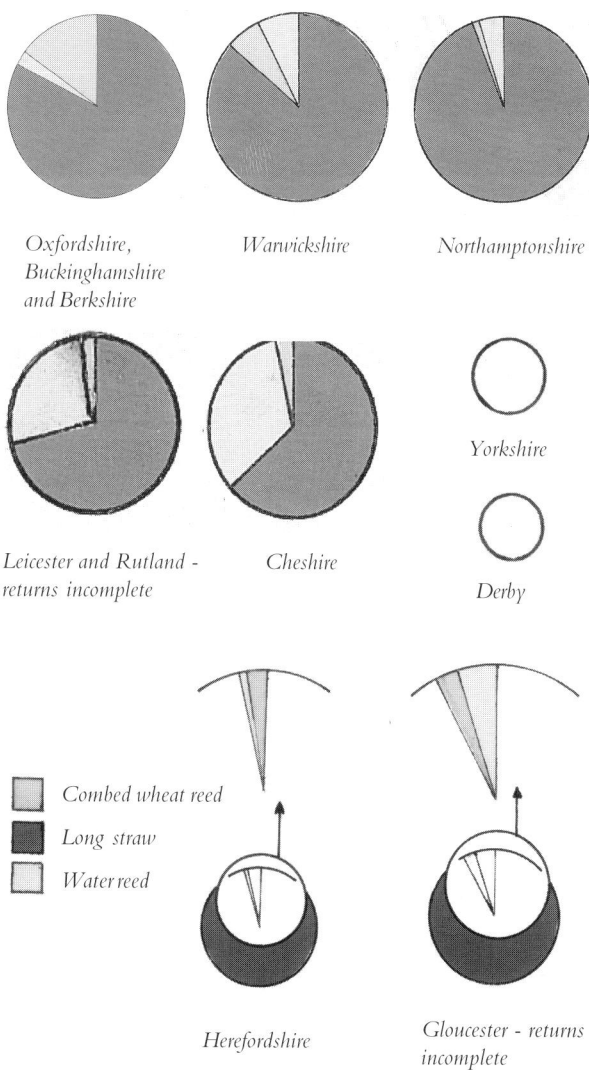

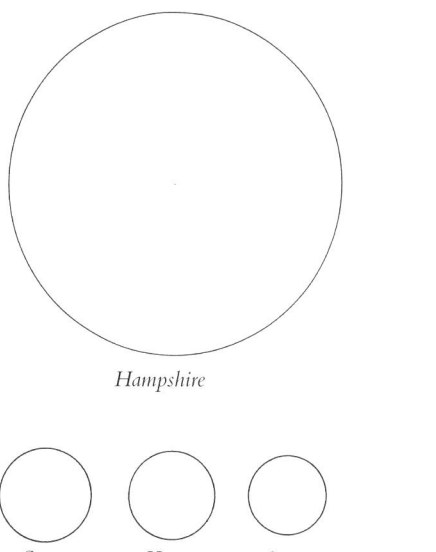

Figure 9 Relative numbers of thatched buildings and types of thatch in the south, from the 1959–60 Rural Industries Bureau survey (Public Record Office 17; see also Table 5 and Figure 8).

Figure 10 Relative numbers of thatched buildings and types of thatch in the Midlands and North, from the 1959–60 Rural Industries Bureau survey (Public Record Office 17; see also Table 5 and Figure 8).

Table 5 Based on a Rural Industries Bureau survey of thatched buildings in various counties, 1959 (Public Record Office 17, see Figure 8).

		Cheshire	Cornwall	Derby	Devon	Dorset	Essex	Gloucester	Hampshire	Hereford	Kent	Leicester & Rutland	Monmouth & Glamorgan	Norfolk	Northamptonshire	Oxfordshire Buckinghamshire Berkshire	Somerset	Suffolk	Surrey	Sussex	Warwickshire	Wiltshire	Yorkshire
Houses in good repair	straw	45	–	–	–	33	2430	155	2601	158	32	–	–	–	340	1753	2	–	–	214	248	2857	–
	combed wheat reed	2	–	–	2925	1069	–	7	–	–	–	–	–	–	2	236	2100	–	–	–	29	–	–
	Norfolk or other marsh reed	28	–	–	–	2	150	–	–	1	10	–	–	736	22	39	5	508	–	–	35	–	–
Houses in poor repair	straw	14	–	–	–	–	520	82	1734	26	14	–	–	118	160	486	1	630	–	21	144	1058	–
	combed wheat reed	–	–	–	2704	1029	–	4	–	–	–	–	–	–	–	14	255	–	–	–	1	–	–
	Norfolk or other marsh reed	–	–	–	–	–	–	–	–	–	1	–	–	91	–	2	–	56	–	–	–	–	–
Barns etc	straw	3	–	–	–	–	515	31	–	–	2	–	–	83	41	1035	–	750	–	48	44	606	–
	combed wheat reed	–	–	–	630	24	–	1	–	–	1	–	–	–	6	10	25	–	–	–	–	–	–
	Norfolk or other marsh reed	1	–	–	–	2	5	3	–	–	4	–	–	274	–	3	–	283	–	–	–	–	–
Other buildings	straw	–	–	–	–	–	3	8	–	1	1	–	–	5	4	29	–	–	–	5	2	25	–
	combed wheat reed	–	–	–	65	8	–	–	–	–	–	–	–	–	–	5	5	49	–	–	3	–	–
	Norfolk or other marsh reed	3	–	–	–	2	2	–	–	1	1	1	–	90	1	1	1	64	–	–	–	–	–
Total		96	450	20	6324	2169	3625	291	4335	194	324	66	130	1683	576	3613	846	4448	232	288	506	4546	30

It was recognised that long waiting lists for thatching meant that thatchers had assured future work and that prices could rise, which was a good thing for the health of the industry. Waiting lists, however, also sorely tried the patience of owners, and risked the replacement of thatch by some other roof covering. In 1961–2 it was noted that in the Midlands, as corrugated iron roofs returned to thatch, water reed, not long straw, was being used for re-thatching, while the roofs of smaller houses, especially in the Vale of Pewsey, Wiltshire were being covered with other materials, because the cost of 'imported' materials (meaning both combed wheat reed and water reed) was so high that it was cheaper to re-roof cottages with other materials (Rural Industries Bureau 1962, 42).

The annual reports of the Bureau begin to analyse the roots of the changes in the industry.

> The increase in the use of the combine harvester has all but severed the direct link between the farmers and thatcher, who has come to rely, to an increasing extent, on the contractor and dealer for the main bulk of his supplies of straw. There are few contractors equipped with reed combers outside Somerset, Dorset and Devonshire and in this area combed wheat reed was readily available. This type of thatch has, however, spread far and wide beyond its native heath in the south west, with a consequence that supplies are strained to a point of scarcity. (Rural Industries Bureau 1962, 42)

The promotion of combed wheat reed (described below) from the south west further east had become the victim of its own success, leading to a scarcity of material. In this context long straw began to look less like a poor relation. In the Midlands and East Anglia, there was an 'encouraging increase' in reasonably-priced long straw, which was also brought into Warwickshire and North Oxfordshire from Lincolnshire, and there had been good local supplies available in Gloucestershire and Worcestershire. New varieties with stems at least 36 inches (0.91 m) in length, *N 59*, *Hybrid 46*, *Jufy I* and *Flamingo*, had been grown widely in the Midlands and East Anglia. These were described by the Bureau as suitable for thatching, but would not be considered so in the 1990s.

The Bureau still had no doubt that Norfolk reed or combed wheat reed were better-performance materials, but seemed less sure that long straw was destined to disappear altogether. In fact, although there was demand for instruction in 'reeding techniques' in the Midlands, 'nevertheless it may be as well for those concerned to bear in mind, in view of the overall picture, that the long straw supply position has shown positive signs of improvement in that area and also in parts of East Anglia' (Rural Industries Bureau 1962, 43).

By 1962–3 the Bureau was noting that the long straw thatchers in the eastern counties were asking for instruction in water reed. In fact thatchers everywhere were asking for instruction in water reed, especially in the Midlands where 'the purchasers of premises originally long straw thatched, and subsequently covered in sheet

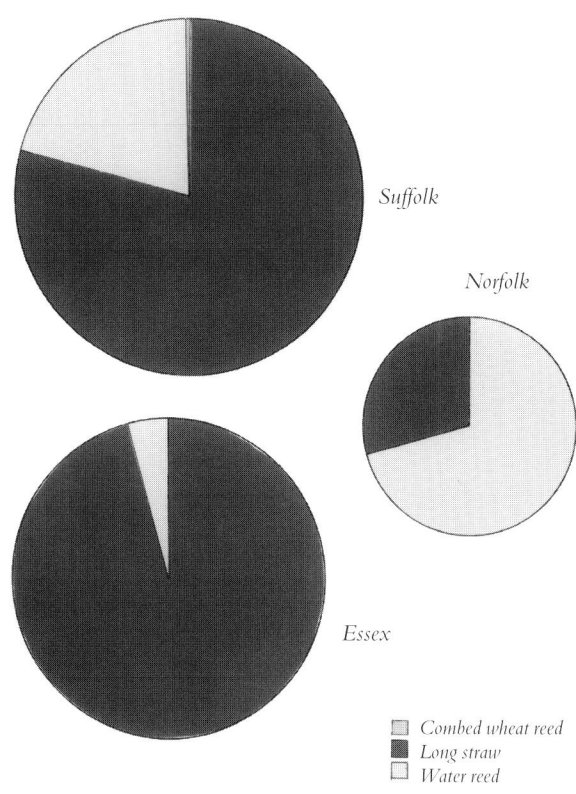

Figure 11 Relative numbers of thatched buildings and types of thatch in East Anglia, from the 1959–60 Rural Industries Bureau survey (Public Record Office 17; see also Table 5 and Figure 8).

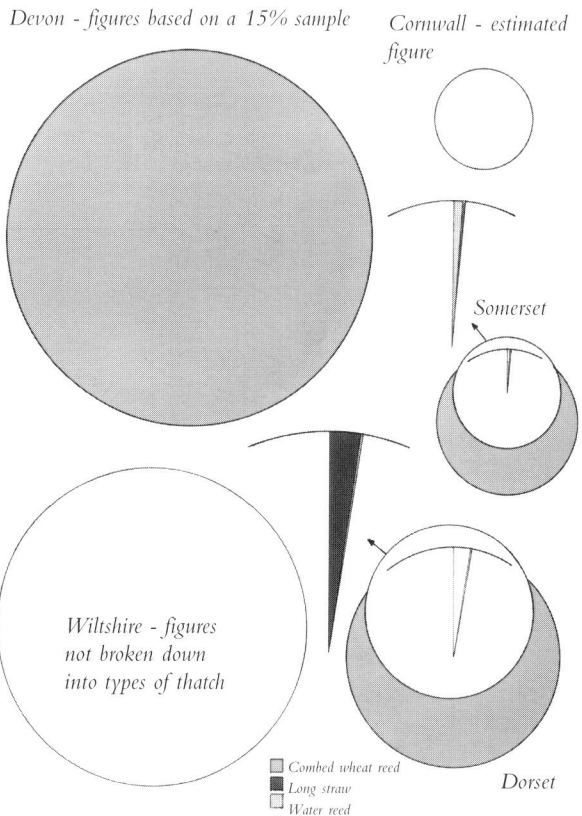

Figure 12 Relative numbers of thatched buildings and types of thatch in the south west, from the 1959–60 Rural Industries Bureau survey (Public Record Office 17; see also Table 5 and Figure 8).

material after stripping, insisted on re-roofing in reed, in preference to straw' (Rural Industries Bureau 1963, 44). Combed wheat reed seemed to have reached its zenith in the Midlands with little call for instruction, 'twenty Midland craftsmen having mastered the method over the past few years' (Rural Industries Bureau 1963). For the first time the annual thatching report acknowledged that a thatching method could be regarded as locally distinctive.

> Rather more time was devoted to long straw instruction, and more properties were thatched in this medium than combed wheat reed which, from the point of view of maintaining the character of a predominantly long straw area, was satisfactory from the purist viewpoint. (Rural Industries Bureau 1963, 42–5)

In 1963–4 the theme of house-owners requesting combed wheat reed is repeated, with requests for instruction in the technique from long straw thatchers in the Eastern counties and the Midlands. However 'a movement towards the raising of the standard in long-straw work developed with a consequent increase in the number of instruction visits in connection with this medium' (Rural Industries Bureau 1964, 58).

By the mid 1960s it was noted that the demand for Norfolk reed was continuing to increase, and both the straws were seen as inferior materials. A year later it was recorded that there were no requests for instruction in long straw. In 1967 the Bureau had shifted its position from regarding combed wheat reed as the material of the future, to water reed.

> Owing to changed methods of harvesting cereal crops, and the consequent trend to grow short straw varieties of wheat, suitable straw for house thatching has become less and less obtainable and the material of the future will undoubtedly be marsh or Norfolk reed. (Rural Industries Bureau 1967, 23)

Pressure from a new kind of house owner was one reason for this. When properties were being restored or changing hands, Norfolk reed was requested. However, there was a change noted in the supply of long straw.

> Although in very short supply in Kent, Surrey, Sussex and Hampshire, an improvement in supplies was noted in both in quantity and quality in the Midland and Eastern counties. A further interesting development is the fact that more thatchers are producing their own long straw. Land which has been purchased or rented by them is being used to grow wheat and then harvested by their own labour force, using harvesting equipment which they have purchased. A variation of this is found where some thatchers, using their own equipment, are harvesting small acreages of wheat grown by their local farmer and taking the straw for thatching as payment for their labour. (Rural Industries Bureau 1968, 19)

To sum up: the Bureau, like new customers for thatch, initially saw many disadvantages to long straw and little to commend it. In terms of agricultural change, it seemed to them more likely to disappear than combed wheat reed, on the assumption that it needed longer straw, which was becoming increasingly rare. It was thought to last less well than combed wheat reed, and many thatchers who used it were thought by the Bureau to be inferior in quality and in need of re-training in another technique. As the quality of long straw thatch rose and it seemed unlikely that it could be comprehensively replaced by combed wheat reed, the Bureau adjusted its opinions and became less critical.

The Rural Industries Bureau's perception of combed wheat reed 1947–68

The 1940s

Clark's 1947 article is enthusiastic about the qualities of combed wheat reed.

> With reed-combed straw available the country builder and architect would have an attractive and endurable alternative to the more costly and rather scarce Norfolk reed.

He describes a demonstration of a reed-combing machine in Devon, clearly impressed by its similarities to the factory production line:

> The sheaves are passed from the rick to a man on the platform and handed on by him (butt ends first every time) to the second man, who lays them on the platform, cuts the string and opens the sheaf. Travelling belts take the straw through the machine, which comes out at the other end as good, clean, unbroken reed with all the flag, grass and heads combed off. This reed, still conveyed by belts, travels down a sloping platform (butt-ends all one way) ready to be tied. (Rubbish and ears drop into the drum and are threshed). (Clark 1947, 448)

This evidence of mechanization, with the potential for perhaps improving machinery and spreading its use, must have been particularly interesting to an organisation with a commitment to progress and modernization. It made it possible to provide hard figures: 'an ordinary day's work with the comber gives between 70 and 80 sacks of threshed wheat and between 900 to 1,000 nitches of reeds'. Clark then gives two estimates for 'Straw Laid Reed-wise' (Figs 9 and 10).

This is the only estimate where a level of regular maintenance is incorporated, not presented as a description of existing practice, but as a desideratum, and with no estimated costing included. This does seem a distortion of what is known to have occurred with all straw thatching. Thatchers who have either a direct or oral history memory of actual practice in the 1930s and 1940s point out that the patching of straw thatch, whether threshed or combed, was so common that 're-thatching', as it would be understood today, was a relatively rare aspect of their work. It would have to be done eventually, although this might be only one pitch of the roof.

The Bureau saw Devon thatching, which was almost exclusively combed wheat reed, as a benchmark. It was considered to be of an exceptionally high quality, and this opinion survived well into the 1950s.

The 1950s

Quality was not just a matter of material, but also of technique, as Williams points out when he describes how the Bureau in the 1950s 'virtually ignored Devon, where the standard of thatching is very high' (Williams 1958, 37). It therefore seemed natural that the technique and material should be chosen as a way of improving the quality of thatch in general.

The Bureau's annual reports reflect the early success of their promotion of combed wheat reed. In 1949–50:

> The combed wheat-reed method of thatching was introduced into some counties which, normally, employed the long straw method, with encouraging results ... Property owners have in many cases been persuaded to use combed wheat straw, thatched reedwise, which is now spreading from its origin in the West Country eastwards into the counties of Wiltshire, Hampshire, Oxfordshire and Berkshire. (Rural Industries Bureau 1950, 22)

Progress had not been entirely smooth and there may have been more optimism than truth in the counties identified in this particular report. There is oral history evidence that combed wheat reed did not arrive in Hampshire until 1955–6 (Jeff King, pers comm) and the Bureau's report of 1956–7 identifies a first appearance of combed wheat reed in that county. There were difficulties encountered in persuading farmers and contractors to make use of the reed-comber and to master its feeding and running but 'its use and value to this type of thatching is now being appreciated in counties where hitherto it has not been used' (Rural Industries Bureau 1950, 22).

In 1952 the Bureau reported a serious shortage of straw and gave figures for the selling price: long straw at £8 and £10 per ton, combed wheat reed between £27 and £32 per ton on site. The transition from long straw to combed wheat continued.

> In order to extend the use of combed wheat reed, which has double the life of ordinary long straw, the Bureau has helped in the introduction of reed-combing machines to those counties where the practice was not previously followed. The Bureau's instructors have spent much of their time in training thatchers in the use of this material, which requires a different technique from long straw thatching. (Rural Industries Bureau 1953, 18)

The following year the Bureau was able to report that the promotion of combed wheat reed was moving it rapidly eastwards.

> The combed wheat straw reed method is normally used in the far west of the area of Devon and Dorset. In these counties, particularly Devon, the method is traditional. Up to the time the Bureau became concerned with the maintenance of standards in the craft, and added two full-time and one part-time thatching instructor to the staff, the combed wheat reed method was scarcely known in other areas ... The policy of introducing this method was continued during the past year, and a few craftsmen in East Anglia have been trained in its use. There is reason to think that the demand for their services may increase, as the few roofs in that area which have been thatched with wheat reed have created much interest. (Rural Industries Bureau 1954, 18–9)

By 1955 the Bureau's discussion of the reasons for promoting combed wheat reed included growing concern that the promotion had got out of hand. There would have been more demonstrations and group instruction in the combed wheat reed technique, but for the shortage of the material.

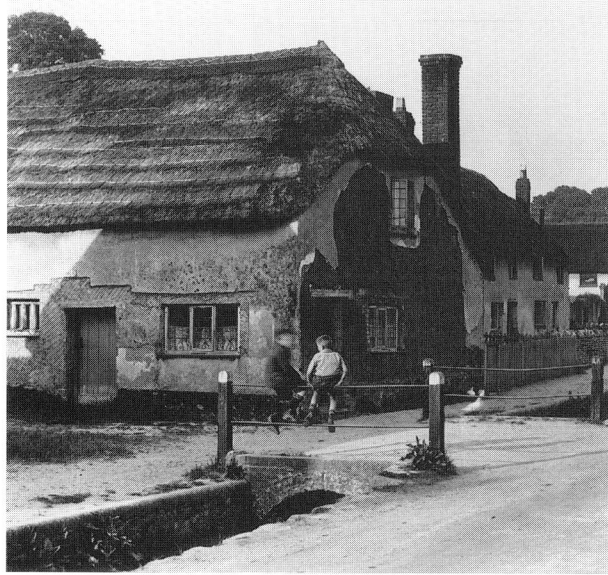

Figure 13 House in Otterton, Devon, 1920s, thatched with combed wheat reed (Devon County Council).

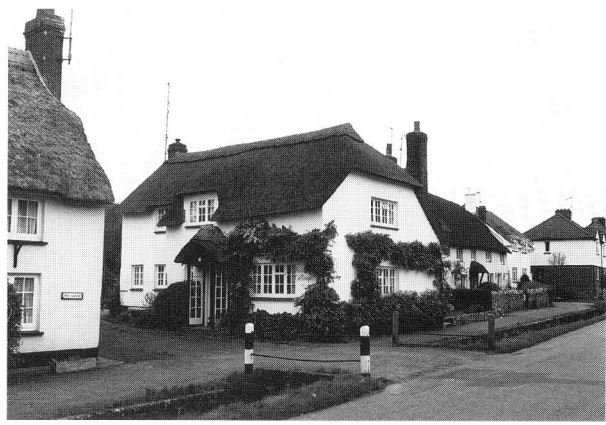

Figure 14 House in Otterton, Devon (same as in Fig 13), rethatched with water reed, 1990. The change in thatching is only one of a number of alterations to the condition and vernacular character and form of the building (Keystone Historic Buildings Consultants).

Until farmers in localities where thatch is prevalent become more aware of the potential market for wheat reed, and are persuaded to install combers, little will be gained by intensifying instruction in combed wheat reed technique. Missionary work among craftsmen has produced a widespread interest in the method, the practice of which, however, must necessarily be limited by the availability of raw material. At present barely enough is produced to meet demands, a state of affairs which demands close liaison between thatchers and farmers. (Rural Industries Bureau 1955, 22–3)

The Bureau failed to predict that its 'missionary' work in popularising combed wheat reed would stretch the sources of supply. The difficulty in localising the production of materials was a crucial one. Most of the supplies of combed wheat reed in the late 1950s continued to come from the south west, where reed combers were manufactured, with attendant transport costs. Quite apart from the philosophical point, an anachronism in the 1950s, that straw thatch ceased to be vernacular when the material was no longer locally supplied, the knock-on effect was a shortage of combed wheat reed, noticed firstly in the 'new' areas, in which (thanks to the Bureau) there was an unprecedented demand, but later affecting the south west too. Devon, where combed wheat was just about universal (the known exceptions being water reed in the Dart valley [Alan Prince, pers comm] and the area around the water reed beds at Slapton Ley) had no other straw tradition to fall back on, and this left a gap that was eventually filled by imported water reed (Figs 13 and 14).

In the 1956–7 annual report the Bureau recorded that it had introduced combed wheat reed into Somerset, Hampshire, Wiltshire, parts of the south Midlands and, more recently, Northamptonshire and parts of East Anglia (Figs 15 and 16). A little more detail was given concerning Northamptonshire.

> Where combed wheat reed thatching has recently been introduced, local authorities and contractors have now been convinced of its soundness and good appearance, and a number of progressive craftsman have been trained in the method.

The shorter varieties and use of the combine harvester made the Bureau believe that:

> the future of straw as a thatching material would therefore seem to be dependent on the installation and full use of reed combing machinery. (Rural Industries Bureau 1957, 27)

In 1959 the thatching instructors were busy producing *The Thatcher's Craft,* the Bureau's pioneering textbook on thatching, first published in 1960. When they got out to work in the field they found a critical shortage of materials with problems finding even enough straw for patching. More attention was focused on water reed than on straw.

The 1960s
The 1961–2 annual report makes it clear that the reed comber had still not spread so far afield as had been hoped. In Somerset, Dorset and Devon there were plenty of contractors with reed combers and combed wheat was readily available. Isolated reed combers existed in Gloucestershire, Suffolk and Northamptonshire and:

> augmented and cheapened supplies of combed wheat reed over limited areas, but until more of these machines are installed in suitable districts serious shortages will persist. (Rural Industries Bureau 1962, 42)

By 1962–3 the price gap in the eastern counties between Norfolk reed and combed wheat reed narrowed. As the difficulties in the supply of combed wheat reed increased, straw thatchers here turned to Norfolk reed. The picture was better in the Midlands, too, due to 'a measure of independence brought about by a few strategically placed reed combers'. In the eastern counties there was only one reed comber, in Suffolk, owned by a master thatcher, Frank Linnet: this eased the problems of demand only a little.

The following year the annual report gave the figures for the numbers of reed combers known to be operating. Within Dorset, Devon, Hampshire and Somerset there were at least two dozen threshing drums with reed-combing attachments. These were augmented by two

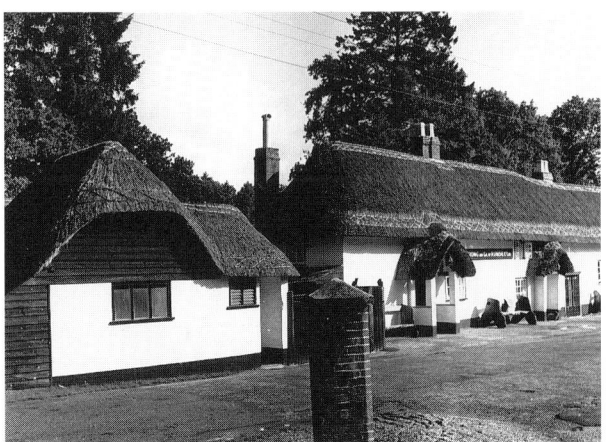

Figure 15 The Sir John Barleycorn, Cadnam, Hampshire, thatched in long straw, 1951 (Victor Schafer/Rural Development Commission).

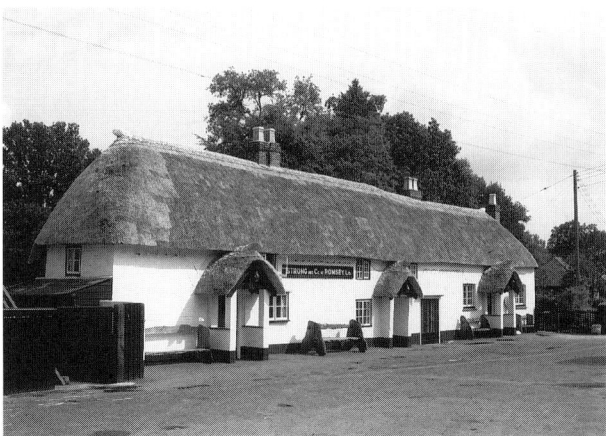

Figure 16 The Sir John Barleycorn, Cadnam, Hampshire, rethatched in combed wheat reed, 1956 (Victor Schafer/Rural Development Commission).

combers in the Midlands and two in the eastern area, which could not supply demand. Transporting combed straw from the south-western counties was inevitable.

The 1964–5 report is largely devoted to the increasing popularity of Norfolk reed, and describes the eastern area as being more in favour of Norfolk reed at the expense of combed wheat reed training. Supplies of wheat straw rose temporarily in spite of lack of quality but towards the end of the period supplies in some areas were almost non-existent. The following year there was a good supply of combed wheat, attributed to a good harvest, although it is clear that by 1968, when the Bureau was replaced by Council for Small Industries in Rural Areas, water reed, both Norfolk and imported, was gradually replacing both the straws.

To sum up: from 1947 until the late 1960s the Bureau actively promoted combed wheat reed, pushing it from the south west into counties where the material and its methods of production had previously been unknown. The reasons for the promotion were connected with the perception of long straw. Combed wheat reed was thought to be more durable than long straw, more visually attractive, and it gave the opportunity to re-train long straw thatchers. In addition, it was taken as read that long straw thatching needed longer straw than combed wheat reed and the increasing use of shorter varieties threatened the prospects for the supply of long straw (Rural Industries Bureau 1957, 31).

The promotion replaced one set of problems with another. Education in method outstripped the availability of combed straw. The principal producers of combed straw remained in the south west, separating thatchers physically from the straw they needed and from their involvement with its production. As demand for the material rose, shortages began to occur. These shortages were additional to the normal shortages resulting from years of poor harvest, and additional to the gradual decline of supplies of long-stemmed straw. It would have been difficult enough dealing with only the latter two reasons for shortages of straw. The Bureau's enthusiastic 'mission' made the situation far worse. Devon thatching in the 1960s was comparatively cheap (Peter Brockett, pers comm). This reflected the relative poverty of that county; the small size of its farms (meaning that it was slow to take to the combine harvester and therefore the production of combed straw held up), and the difference between the relatively small numbers of incomer owners in Devon as compared to rural regions in the Midlands and south east, within striking distance of urban areas.

Suppliers of combed wheat reed in Devon could get a better price by selling it out of the south west, leaving a scarcity on their own doorsteps. As the results of shortage began to show, water reed, imported from abroad, began to fill the gap. There were other important factors. Water reed had a deserved reputation for longevity close to home in East Anglia, and the additional expense of using the material became less and less of an issue as owners with more disposable

---

R. W. FARMAN
North Walsham, Norfolk

Established Many Centuries

NORFOLK REED THATCHING

Work carried out in any part of the country by Norfolk Experts.

Estimates given free on receipt of plans

Reed Thatching can now be treated with a Fire-Proofing Solution at small extra cost.

Existing Straw-thatched buildings can be re-thatched with Reeds.

Advice given on application

Repairs to existing Reed-thatched Roofs carried out expeditiously, as our Thatchers are working in many districts at all times.

Telephone 76.   Telegrams: "*Farman, North Walsham*"

Figure 17 An advertisement for the Farman firm in Laxton's Price Book, *1938, giving an idea of how sophisticated the leading Norfolk firm was at that date.*

income than their agricultural predecessors moved into rural areas.

The Rural Industries Bureau's perception of water reed 1947–68

The 1950s

The Bureau originally decided not to intervene in the water reed thatching industry, perceiving it to be healthy and independent from straw thatching. Norfolk reed thatchers, unlike straw thatchers, were not anchored to farms. The fact that they purchased their material in ways more like any other building contractor had liberating effects on business. The Cowells of Soham were sending their thatchers considerable distances before the turn of the century and, before the Second World War, one famous firm, Farman Brothers of North Walsham, were travelling all over the country, using Norfolk reed delivered by train (Fig 17). They worked as far afield as the Devon/Dorset border in the 1930s (Ward 1939). Farmans were so well-known, and so widely travelled that there was a sense in which for 'Norfolk reed' in the 1940s, read Farmans. It seems to have taken the Bureau some time to realise that there were other, smaller, more local firms using Norfolk reed (and reed sourced in Kent, Suffolk and Dorset, among other places), and it was not long before they became involved. Throughout the annual reports the

Bureau reiterated its view about the durability of water reed. A sentence from the 1954–5 report is typical.

> In the hands of a master craftsman, it produces the most durable of all thatch, and while a roof thatched with reed will normally cost more, size for size, than one thatched with straw, many consider its use an economy as it has a far longer life. (Rural Industries Bureau 1955, 22)

Before 1951–2, there is no mention of water reed thatching in the annual reports, which were designed for public consumption. In 1951, however, the Quarterly Report on the Work of Crafts Section, an internal document, hints at the reasons for the inflexible supply of Norfolk reed in the early 1950s.

> The supply of Norfolk reed remains unsatisfactory: there does not appear to be any surplus reeds for sale outside the ring of regular buyers and users. (Public Record Office 9)

It is difficult to know whether the Bureau meant restrictive practices by this reference to a 'ring'. Some Norfolk reed thatchers resented the spread of 'their' material further afield. An article in the *Guardian* quoted a Norfolk reed thatcher complaining about the depredations of thatchers from elsewhere.

> They comes and sneaks it away under our noses ... By rights they oughter keep to straw, what God made their natural material. (Cohen nd)

In the 1951–2 Annual Report, the Bureau is more discreet, and refers to problems with the Norfolk reed harvest as a result of poor weather and shortage of reed-cutting labour for a task which was 'both arduous and unattractive'. The Bureau started to look for other sources of English water reed and reported that 'some reeds in Hampshire have been harvested and several roofs thatched without difficulty in this material'. The location was almost certainly the Southampton and Christchurch areas, mentioned in the 1952–3 report, which gave the life of Norfolk reed as 'between forty and sixty years' and noted that there had been a shortage for some years. The Hampshire reeds were found to be less strong than Norfolk, but it was understood, on the advice of an expert Norfolk reed-cutter, that they might improve with regular harvesting. The Bureau saw harvesting as an alternative occupation for fishermen and planted some Norfolk reed at Christchurch to compare it with the local reed.

In 1953–4 the Bureau stated that 'apart from Norfolk and neighbouring counties where Norfolk reed is almost exclusively employed, straw threshed in the normal way or combed ... is the principal material used' (Rural Industries Bureau 1954, 18). The following year, this was slightly refined with a reference to straw being 'almost universal' in the south of East Anglia. By 1953–4 some East Anglian craftsmen had been trained in use of combed wheat reed. It seems likely that these were water reed thatchers, as the report notes that the supply of Norfolk reed had been badly affected by floods and water reed thatchers were seeking alternative materials.

As with the other materials, the Bureau gradually developed a broader understanding of the supply and husbandry of water reed. Norfolk was producing less because 'scythe men have dwindled in number' (Rural Industries Bureau 1954, 19). By 1953–4 the Bureau was collaborating with landowners and engineers to test

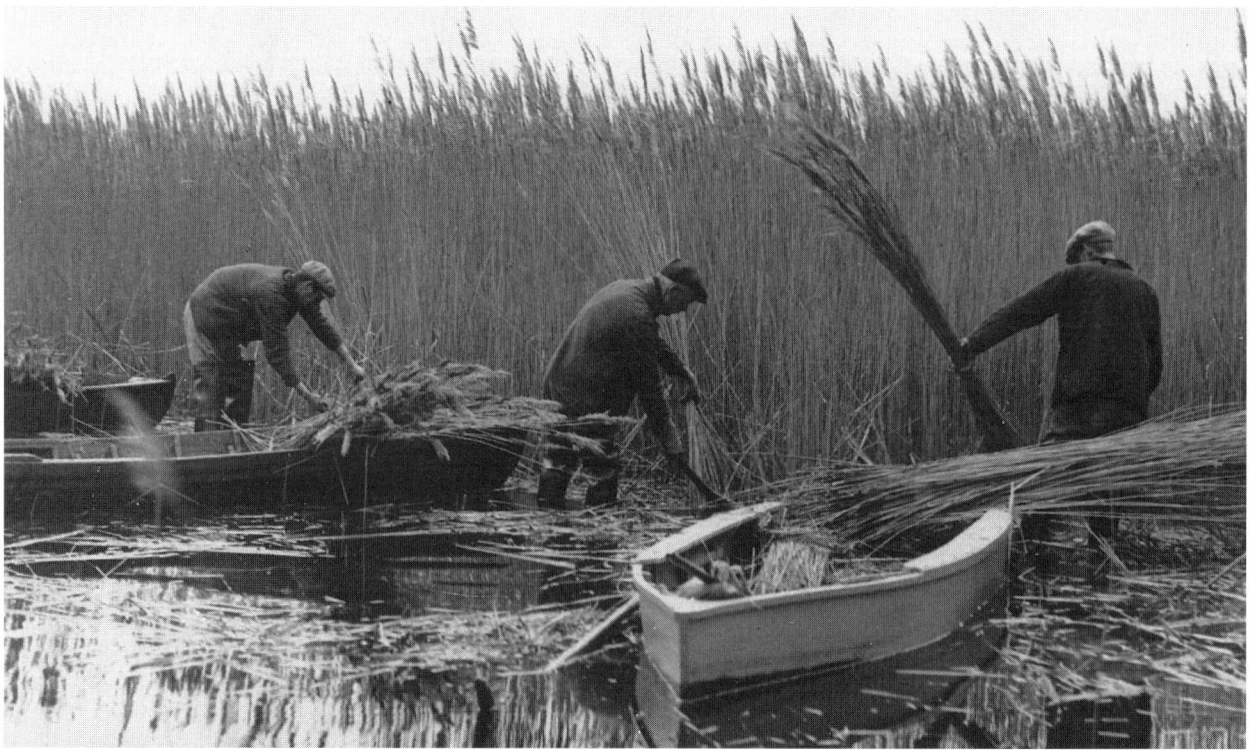

Figure 18 Reed cutting by hand, c1934, Norfolk. Norfolk Studies Library, Accession number 5975 Class R (Eastern Counties Newspapers).

mechanical means of cutting: 'a number of machines were tested in early December at the Hickling reed beds' and a motorised scythe was tested on reed beds in Devon (presumed to be Slapton Ley), but further experiments were needed. It was noted that the reed from the Christchurch reed beds was improving 'due to regular annual harvesting, amounting to cultivation, over the past five years' and provided Hampshire thatchers with a limited amount of material.

In 1954–5 the Norfolk water reed harvest was poor again, and the Southampton and Christchurch experiment had to be abandoned 'owing to a number of factors outside the Bureau's control' (Rural Industries Bureau 1955, 24). The experiment had been transferred to an area near Keyhaven, on the Solent, where mechanical cutters (including a prototype of a design by the Bureau), had been tested. The following year, 1956–7, the harvest was no better but there were continuing attempts to find a satisfactory mechanical reed cutter and one of the Bureau's own designs for a scythe was being manufactured in Wiltshire and had been demonstrated twice in Norfolk and once in Essex, Worcestershire, Suffolk and Dorset.

The problems of the supply of Norfolk reed were complex. The shortage of (human) reed cutters was one (Fig 18). Seasonal problems ranged from flooding, which delayed cutting and could rot the reed, to damage from early frosts, which spoiled the young shoots. There was, too, an apparent lack of enthusiasm among the reed producers to increase production to meet demand. The Bureau found this difficult to understand but continued its search for supplies to supplement Norfolk reed, particularly sources further west. This, it was hoped, would cut down transport costs. By 1958–9 a small number of thatchers had been encouraged to rent reed beds 'reasonably adjacent to their sphere of operations and to harvest reed for themselves' (Rural Industries Bureau 1959, 28). A small business had been established in Hampshire, with cutting rights to a number of Hampshire reed beds and employing twenty cutters. A 1960 article in *The Field* (Whitlock 1960) gives a little more information about this enterprise, started by a Southampton businessman, C Block. In the 1958–9 season 15,000 bundles were produced from Keyhaven, Totton, Lymington and Christchurch (Hampshire) and Radipole Lake (Dorset).

> Reed beds in Dorset and Glamorganshire, developed and administered by a municipal authority and a landowner respectively, have also added their quota to help alleviate the shortage. (Rural Industries Bureau 1959, 28)

It was also reported that a Devonshire business man had experimented with importing reed from Holland (this was not necessarily Dutch-grown reed but may have simply have been imported via Holland). This was expensive but apparently shipped internally and used by thatchers within striking distance of ports of entry. The Bureau hoped that competition from imports would stimulate the Norfolk growers to 'utilise to the full their own vast acreage' (Rural Industries Bureau 1959, 29).

In 1959–60 their hopes seem to have been fulfilled in part, at least, and at the same time as reed from the Hampshire saltings was being harvested and Dutch imports were being used, Norfolk production was increased.

> Reed from Norfolk was also obtainable in greater abundance than of former years, cutters having shown a certain keenness during the 1958–1959 winter to cut a greater acreage of both reed and sedge for ridging. This increased and most welcome interest may have been due in part to a number of timely well-written articles and press reports. (Rural Industries Bureau 1960, 33)

Requests for instruction in water reed (and combed wheat reed) were received from the Eastern and Midland areas from long straw thatchers.

The 1960s
By 1960–61 water reed was clearly gaining popularity and requests for tuition were received from the Eastern region, the Midlands, the Southern area and 'even in Devonshire, the heart of the combed wheat reed country' (Rural Industries Bureau 1961, 30). When Alan Fooks first encountered water reed in what had been a long straw area of Dorset it was 'like finding a bull in the kitchen' (Alan Fooks, pers comm). Undaunted, he decided to learn the method and asked for instruction from the Bureau and Reg Clarke came down from Winchester to teach him.

In 1961–2 flooding had limited the cutting season in Norfolk and it was reported that thatchers with Norfolk reed commissions had made use of the imports from Holland, where orders were promptly fulfilled. The Bureau was puzzled by the 'paradoxical situation', where the reputation of Norfolk reed was simply boosting the level of imports from foreign sources, instead of encouraging Norfolk reed growers to harvest a larger proportion of the marshes.

In 1962–3 the 'Norfolk reed technique' seemed to be attracting interest everywhere, but especially in the traditional long straw areas. As with combed wheat reed, the limits were largely that of the supply of materials: 'had supplies been adequate requests for instruction could have been implemented with greater success' (Rural Industries Bureau 1963, 45). Imported Dutch reed was used for at least 25% of the 'reeding' jobs (this seems to mean water reed, rather than water and combed wheat reed). In the south it proved increasingly difficult to obtain good water reed, whatever the provenance. For the first time there was a suggestion that imported water reed was not always of the first quality for thatching.

> Reed from Holland rose to an uneconomic price, but was perforce used, irrespective of quality when the reasonably priced deliveries from Norfolk became difficult, if not impossible to obtain. (Rural Industries Bureau 1963, 46)

The Bureau was diligent in efforts to understand water reed husbandry and the problems of cutting. Their intervention, encouraging the development of water reed beds elsewhere and the development of mechanical cutters, was pragmatic and sensible. What obviously puzzled them was the apparent failure of the Norfolk reed industry to respond to demand: 'the question as to why the vast reed-bearing beds of Norfolk do not produce a greater harvest is one to which no one will readily hazard an answer' (Rural Industries Bureau 1963, 46). This seems to have been a low point, during the Bureau's existence, in the supply of indigenous water reed, just before the reed-growers began to respond to the increased demand.

In 1963–64 the Norfolk reed harvest was far better than it had been for some time. It was reported that half of the harvest was mechanized and water reed was abundant, although the price rose as the season advanced and increased transport charges pushed the price to thatchers up further. The availability of reed meant that there was little need for imported Dutch reed in East Anglia or the Midlands.

By 1964–65 the Bureau was convinced that water reed, not combed wheat reed was the material of the future.

> This swing away from straw, which originated in the need for an alternative to offset its scarcity, has been further influenced by the realisation that Norfolk reed, although initially more expensive, is economically a better long term proposition, because of its durability. (Rural Industries Bureau 1965, 18)

A Siega Reed Harvester was imported from Denmark with the expectation that this would improve the quantity of supplies.

In 1967–8 the Bureau reported that courses in Norfolk reed thatching techniques had been held at Knuston Hall, Northamptonshire, during the winter months.

To sum up: the Bureau saw water reed as the best of the three common thatching materials. It may be significant that Fred Cooper, one of the most active thatching instructors was 'a Norfolk Reed man'. The principal problem with water reed was its supply. Had it been more readily available countrywide there is little doubt that the Bureau would have been promoting water reed, rather than combed wheat reed, with missionary zeal.

Conclusion

The Bureau's annual reports on the changing use of materials from 1940–68 were produced from the perspective of the only organisation with an overview of thatching in the period. Their investigation of the nature of demand and supply are crucial to an understanding of the twentieth-century history of thatching.

Initially, water reed (and this meant Norfolk reed) house thatching was seen as different in kind from the two straw methods, long straw and combed wheat reed house thatching. Norfolk reed, according to the Bureau, was largely limited to Norfolk and parts of the adjacent counties. Although long straw is the term consistently used in the annual reports, a 1951 account from a former trainee thatcher calls it 'short and long tail wheat straw' (Appendix 2). In spite of Cosmo Clark's article in *Agriculture*, the thatching officers who worked for the Bureau seem to have had no trouble in identifying and naming long straw as a tradition, although there is absolutely no mention in the reports of variation in ridges or eaves and verges detailing or the consequences to appearance and performance of different methods of ricking and drawing, which are described by thatchers who worked in the 1940s. However much variety of technique and detail there was in long straw, and however diverse its route and processing from field to roof, it could be, and was, considered by the Bureau as one type of thatch found in the 1940s from Hampshire to Kent.

If the thatching officers noted local tradition within the thatches, as no doubt they did, little appears in the annual thatching reports. The following reference is remarkable for its rarity:

> The technique of thatching varies little from one district to another, apart from local styles of finish, such as the up-pointed gable-end typical of Suffolk, made by bunching up the sheaves, or the rounded-off one common in Wilts (Public Record Office 3, 60).

In the early 1940s combed wheat reed, as far as the Bureau was concerned, was found exclusively in Devon, Dorset and Somerset, and was ubiquitous in Devon. Again the Bureau made no distinction between styles or various local details of combed wheat reed or the ways these were amended by thatchers as they travelled east.

Of course there is plenty of evidence to show that the Bureau's perception was a relatively crude one. Norfolk reed had spread far beyond Norfolk long before 1940 although caution must be exercised in assuming that the travelling Norfolk reed firms were able to thatch large numbers of buildings in counties far from home. There were other sources of water reed pre-dating Bureau intervention, Abbotsbury in Dorset, for example, and pockets of water reed thatch close to these sources and on the estates that had access to them.

The Bureau's missionary work in spreading combed wheat reed eastward into the areas where long straw had existed in 1940 needs to be put into the conservation context of the time. The thatching instructors had a commitment to the re-establishment, maintenance and extension of the market for thatch at the expense of other roof coverings (Fig 19). There was no impetus, or brief to investigate and/or preserve regional variations in either thatching materials or techniques (beyond what was required for good performance). In the 1940s, 1950s and 1960s there was little impetus to this end from outside the Bureau either. In spite of pioneering research by Innocent (1916), in 1940 there was not a widespread interest on the part of building historians in the finer points of rural vernacular buildings. Even the interest and understanding of vernacular building materials that had developed by 1968 cannot be said to have included a good understanding of thatch and had not been absorbed, to any great extent, by the government organisations

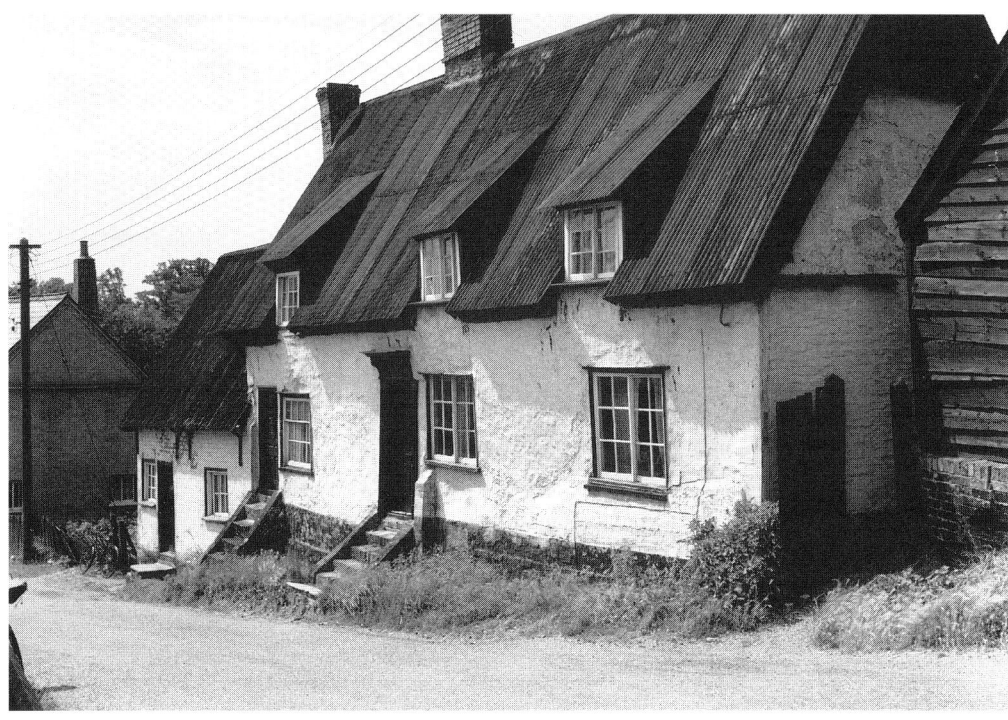

Figure 19 A house reroofed in corrugated iron, replacing thatch, Elesworth, Cambridgeshire. The Rural Industries Bureau described this picture as an 'Aweful Warning'. (Victor Schafer/Rural Development Commission)

with an interest in building conservation, or by many of the incomers who were buying traditional rural houses.

The choice of combed wheat reed as the preferred thatch in the 1940s and 1950s was obvious. In the eyes of the Bureau there were only three significant thatches to choose from anyway, and it was unthinkable, given the road system and transport costs in the 1940s and 1950s and the limited availability of Norfolk reed, that water reed could be pushed throughout England. It was not until the 1960s, when it appeared that foreign imports might be able to make up the deficit that it seemed possible that water reed might become ubiquitous. Of the two straw thatches, combed wheat reed, for the reasons outlined above, was seen as more durable than long straw, more likely to survive within contemporary agricultural practice. Although this is not mentioned in the annual reports, combed wheat reed was also less bulky and therefore cheaper to transport than long straw. Wheat reed also had the attraction of involving machinery special to thatching, the reed comber, and this offered prospects of improving that equipment in a progressive way.

The question of precisely how much impact the Bureau had on changing materials is impossible to judge, but it can reasonably be said to have been a large one. It is unlikely that combed wheat reed would have travelled so far so fast outside the south west without the Bureau's help. Even with active promotion from a centralized body, its progress east and north was disappointing. Nevertheless it effectively began to squeeze long straw out of the Midlands and the southern counties. The spread of water reed, especially from foreign sources, would certainly have been a great deal slower without the Bureau's enthusiasm for combed wheat reed. The demand for combed wheat reed from the Midlands and the southern region led to a scarcity, and this, in turn, opened up a market for water reed in both the south west and in counties without a coast, hastened by the Bureau's instruction in the technique. Arguably, even without the Bureau's intervention, water reed, including imports, would have shifted into the traditional straw areas anyway (Fig 20), as it did in the Netherlands, where water reed imports have a long history and have gone hand in hand with sophisticated marketing campaigns.

In 1965 the Bureau conducted a census of thatchers, identifying their numbers, ages, status and ability to fit different types of thatch (Table 6). It is not clear how the

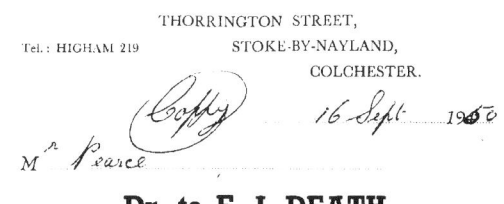

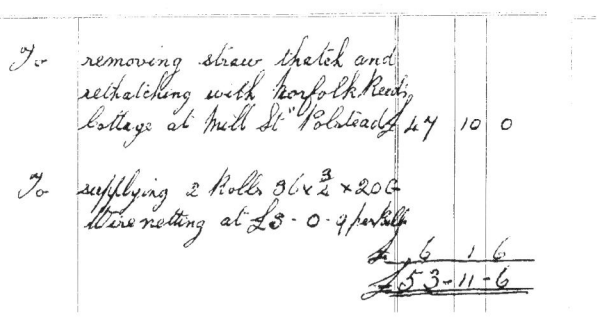

Figure 20 Straw (long straw) replaced with water reed on a cottage in Polstead, Suffolk, in 1950. From the personal papers of the Death family.

Table 6 Rural Industries Bureau Census of Thatchers, 1965, showing competence in different materials as well as status and age range.

County	total numbers	long straw	combed wheat reed	Norfolk reed	age 15–20	age 21–35	age 36–50	age 51–65	age over 65	master-men	apprentices	trainees	employees
Bedfordshire	16	15	7	3	2	1	3	6	-	11	2	-	3
Berkshire	17	16	10	9	3	4	4	5	1	15	1	1	-
Buckinghamshire	9	8	4	3	1	6	1	1	-	9	-	-	-
Cambridge	22	22	-	10	1	-	2	7	6	15	1	-	6
Cheshire	1	1	-	-	-	-	-	-	-	1	-	-	-
Cornwall	6	-	6	-	-	-	1	4	1	6	-	-	-
Derbyshire	4	4	1	-	-	1	2	1	-	4	-	-	-
Devonshire	96	-	96	32	6	8	25	40	17	83	3	5	5
Dorset	58	58	57	20	4	11	8	21	14	41	1	2	14
Essex	40	38	8	9	2	3	10	16	-	34	2	-	4
Glamorgan	3	3	2	2	-	1	1	1	-	3	-	-	-
Gloucestershire	8	8	3	5	2	2	1	2	1	6	1	1	-
Hampshire	68	68	38	24	4	7	18	21	17	57	1	3	7
Hertfordshire	5	4	1	1	-	-	1	2	1	4	-	-	1
Huntingdonshire	15	15	12	15	-	1	1	4	-	6	-	-	9
Isle of Ely	5	4	-	4	-	-	-	4	-	4	-	-	1
Kent	6	4	1	1	1	-	-	4	-	5	-	-	1
Kesteven	1	1	-	-	-	-	-	-	-	1	-	-	-
Lancashire	1	1	-	1	-	-	1	-	-	1	-	-	-
Leicestershire	2	2	-	-	-	-	1	1	-	2	-	-	-
Lindsay	1	-	-	1	-	-	-	-	-	1	-	-	-
Middlesex	1	-	1	1	-	-	1	-	-	1	-	-	-
Norfolk	45	20	2	35	-	3	3	9	4	17	-	-	28
Northamptonshire	27	27	13	13	2	2	4	10	2	19	2	-	6
Nottinghamshire	1	1	-	-	-	-	-	-	1	1	-	-	-
Oxfordshire	25	25	11	4	1	4	8	11	1	22	-	3	-
Rutland	3	2	-	2	1	-	2	-	-	2	1	-	-
Shropshire	1	1	-	-	-	-	-	1	-	1	-	-	-
Somerset	27	-	27	8	4	4	4	10	5	21	-	5	1
Staffordshire	1	1	-	-	-	-	-	-	-	1	-	-	-
Suffolk	56	52	10	22	3	-	13	20	7	44	2	-	10
Surrey	1	-	-	1	-	-	1	-	-	1	-	-	-
Sussex	16	16	2	10	-	4	3	8	1	16	-	-	-
Warwickshire	11	10	8	3	-	5	1	4	1	11	-	-	-
Wiltshire	57	57	46	8	5	8	13	26	5	40	1	6	10
Worcestershire	8	8	3	3	-	2	4	2	-	8	-	-	-
totals	**664**	**492**	**369**	**250**	**42**	**77**	**137**	**241**	**85**	**514**	**18**	**26**	**106**

information was gathered and no doubt there were imperfections in this, which may explain some of the apparent oddities in the census figures. Nevertheless, the relationship between the numbers of thatchers and their ability to thatch, say, long straw, is probably a fairly good guide to the strength of that tradition in any one county. All 27 thatchers in Northamptonshire, for example, were capable of thatching in long straw. Thirteen could also thatch combed wheat reed and 13 (possibly the same 13?) could thatch in Norfolk reed. Some of those who could thatch combed wheat reed had undoubtedly been cross-trained by the Bureau. In Dorset, a county in which the Bureau's 1960 survey had identified very little long straw, all 58 identified thatchers were capable of thatching it, including four thatchers who were under 20 years of age, which seems surprising. All but one were also capable of thatching combed wheat reed. In Somerset, where a tiny amount of long straw had been identified in 1960, there was no longer a single thatcher capable of fitting it.

The Bureau never set out to be responsible for obliterating one, or any, of the three thatches in which it took an interest. It tacitly accepted that much of what was happening to thatching was the result of factors outside its control, either the presence of 'poor' long straw thatchers in the 1940s or, later, agricultural changes which made straw difficult to obtain. On the other hand, there were times when there was a need to explain the nature of intervention in the annual reports, as it became clear that the promotion of combed wheat reed was having unforeseen consequences.

> It is not intended to push the combed wheat reed method of thatching to the exclusion of the other styles. The instructors are skilled in all three methods and a large proportion of their time is given to advising and instructing many scores of craftsmen who are using and will probably continue to use exclusively either long straw or Norfolk reed. (Rural Industries Bureau 1954, 19)

Six years later, the picture looked quite different, and it seemed safer, for the sake of thatchers, to ensure that all were cross-trained in all three thatching methods. There may have been a element here of the Bureau rationalising a situation which had come to exist, and perhaps a desire to gloss over a certain embarrassment about its own contribution to the supply problems of materials.

> The shortage and uncertainty of supply of habitually used materials has forced upon thatchers a more complete realisation that no longer can they rest secure possessed of a specialised knowledge of one thatching method, but must become masters of all three techniques and be

prepared to switch from one to another as the customer, or, more often, necessity, demands. (Rural Industries Bureau 1961, 30)

In 1965 the Rural Industries Bureau produced an extensive report on thatching (reproduced here as Appendix 1), 'conducted to assess the value of thatching as a national asset' and covering the state of the craft at the time and the factors affecting it. Their 1965 census of thatchers was probably produced for this report. After considering the economic, historic, aesthetic and architectural value of thatch, the Bureau assessed 'the property situation' and the 'materials situation'.

'The Material Situation', written by Fred Cooper, is of especial interest. The report was designed to have a wider circulation than the annual thatching reports, which were produced side-by-side with reports on other rural industries. In it the Bureau was ready to predict the demise of straw thatch altogether. Cooper's contribution included a rehearsal of the problems associated with the supply of materials, and an assumption of a transition from straw to water reed, at this date understood to be Norfolk reed.

> 31. Only careful observers and those having access to appropriate records will be aware of this transition, and some reasons for it are given in the following paragraphs. (Rural Industries Bureau 1965)
> A few years ago it was an easy task to indicate clearly defined areas where one particular thatching material was used predominantly and even exclusively. For example it was possible to follow a belt of long straw running westerly from Chelmsford in Essex, through Hertfordshire and Bedfordshire, and thence through Northamptonshire and parts of Oxfordshire, north west to Worcestershire and Herefordshire.
>
> 28. Throughout such an area, to which might be added the Suffolk/Norfolk border and other similar pockets in Hampshire and Wiltshire, property owners and thatchers thought mainly in terms of long straw thatching and relied on the local farmer for a supply of good straw grown from a recommended variety of wheat suitably harvested with a self-binder and threshed with a threshing machine.
>
> 29. Combed wheat reed would have been predominant in Devon, Somerset and Dorset / any other material [in these areas] would have been regarded as an intrusion upon tradition.
>
> 30. Norfolk Reed
> The situation described in (28) and (29) above no longer pertains, the trend now being towards the more extensive use of the much more durable Norfolk reed, regardless of location or tradition. (ibid)

Cooper saw two reasons for the changes he outlined. Firstly, rural properties had changed hands and the new kind of property owner made different demands of a thatch. Architects, builders and property owners 'have been looking for a material more durable than long straw or combed wheat reed' (op cit, paragraph 32). Cosmo Clark's insistence in 1947 that thatch should and could be a roofing material acceptable outside the farm had been justified, but it seemed that this could only be sustained by abandoning the material that was still obviously 'farmed'.

Agricultural practice had reduced the supply and quality of straw. Cooper described the problem of the replacement (with some exceptions) of the self-binder and threshing machine by the combine. He also noted an experimental use of a growth regulator, 'Cycocel', in Suffolk, which reduced the straw by six or seven inches, to assist combining. All in all, new technology and new techniques in wheat-breeding and growing made the prospects for either of the straw techniques look pretty miserable.

The conclusion in 1965 was that by 1975 Norfolk reed would 'become the main medium for thatching'. This introduced some remarks about the difficulties in producing enough of it to satisfy the needs of the 600 thatchers in England known to be trading at the time. If the numbers of thatchers were to be maintained at the 1965 level, the estimated annual consumption of water reed was expected to be 2,300,000 bunches. However 'if nothing can be done to stimulate supplies of other mediums and thatching becomes dependent entirely on marsh reed, 6,000,000 bunches will be required'. If Cooper suspected or knew that this increase could only be accommodated by foreign imports, it was not spelled out. It may have been politically awkward to admit that an industry developed in order to provide employment in rural areas might find itself dependent on foreign enterprises.

THE COUNCIL FOR SMALL INDUSTRIES IN RURAL AREAS AND THE RURAL DEVELOPMENT COMMISSION

By 1966 it seemed that the whole system of Rural Industries Organisations was in some disarray. There were three, more or less autonomous, arms to the system. 'Technical advice' was provided by the Bureau (which included the thatching instructors); there were also 'the field organisations', including the rural community councils and organisers, who were sometimes quite closely involved with the local Master Thatchers' Association, and attended meetings; and there was 'the loan fund', designed to help rural businessmen to buy equipment, workshops etc. There was not always an acceptable level of respect or cooperation between the three sections and it was agreed that what was needed was 'a unified, controlling, executive body for the rural industries service' (Public Record Office 10). The Bureau was replaced by the Council for Small Industries in Rural Areas which, in its turn, was subsumed by the Rural Development Commission in 1988.

From 1968 the Council for Small Industries in Rural Areas and the Rural Development Commission built on the foundations established by the Rural Industries Bureau. Retired thatching instructors who experienced the change from one agency to another say that it made little difference to their activities. They concentrated on training (described in detail in Chapter 6 below) and on the vexed question of the supply of materials. To this end, the Rural Development Commission pressed for research projects (some of which they funded) that might help to solve specific problems; the 'rapid decay' problem that affected combed wheat reed in the south west in the 1970s, for instance, and the problem of fire risk. These research projects have been of immense value to the thatching industry as the building blocks on which subsequent research has been erected as well as drawing attention to the difficulties under which thatch labours as a modern roofing material.

The late Peter Brockett, a thatching instructor with the Council for Small Industries in Rural Areas and latterly an independent thatch consultant, was unflagging in the pursuit of a broad range of research directed to the improvement of the performance of thatch, the reduction of fire risk, the improvement of thatching standards and in disseminating the results of research to thatchers. Information resulting from the University of Bath research (Kirby & Rayner 1989) included the findings that tight packing of thatch on the roof could speed up decay. This was a matter of technique that needed to be communicated to thatchers, and Dr Joe Kirby, working on the project, was encouraged by Peter Brockett to attend meetings of the Regional Master Thatchers' Associations and communicate his findings. This is one example of the value of professional associations as channels of communication between research and working thatchers.

5 Trade organisations

THE MASTER THATCHERS' ASSOCIATIONS

The regional associations were the earliest organisations, and were initiated by the Rural Industries Bureau in 1947–8.

> The Bureau, having found the thatchers, endeavoured to bring them together as a fraternity in local Associations, which could speak with authority on behalf of the craft in the locality. (Rural Industries Bureau 1948, 26)

After the difficulties experienced with the Ministry of Labour in 1947 in setting up a Vocational Training Scheme for an industry without a trade organisation, the Bureau was keen to see one develop. The decision to sow the seed with regional organisations was a response to the realities of the industry. The Bureau's survey of the late 1930s had revealed massive variation in quality and prices within regions as well as from region to region and this was likely to be corrected first at a local level. The Vocational Training Scheme was local, not centralized. Self-help amongst thatchers also made sense locally, where individuals shared the same problems of sources of supply.

The minutes of the preliminary meeting of the Somerset Master Thatchers' Association give a flavour of the first efforts at trade organisation in one county. The meeting was held at St Margaret's, Hamilton Road, Taunton on Saturday 31 January, 1948: 'to consider the formation of a Master Thatchers' Association for the County'. Colonel G A Garton (Chairman of the Somerset Rural Community Council) took the chair and there were 23 others present including Commander H T Andrew. It seems that 18 of the 23 were thatchers, the others were presumably officials like the Rural Industries Organiser. Apologies were received from F Davies, the thatching officer of the Rural Industries Bureau, from Mr Reike of the National Trust and Mr Harris, the County Architect (Somerset Master Thatchers Association Archive).

There were nine purposes to the organisation:

1. to establish the proper status of the thatching trade
2. to formulate a reliable standard in the quality of work, and to ensure that, as far as possible, work should be carried out only by those competent to do it
3. to adopt a schedule of prices for the various grades of the thatchers' craft
4. to control the training of applicants intending to become qualified thatchers
5. to be responsible for negotiating wage rates
6. to work for the improvement of the quality of thatching materials
7. to organise, where required, the group purchase of materials
8. to safeguard the interests of individual members, eg by special insurance, legal defence etc
9. to do all such things as may seem incidental or conducive to the attainment of the above purposes or any of them.

In the discussion that followed, thatchers expressed doubts about the financial incentives to take a trainee on the Vocational Scheme because of loss of time in wet weather. Mr Webber of Dunster felt that a thatcher could make spars and gads in wet weather. There was a problem obtaining supplies of spars in Somerset, although there was said to be no shortage in Dorset. Oxfordshire and Buckinghamshire's Master Thatchers' Associations were cited as having helped to resolve problems in the supply of materials. Mr Edwards expressed a reluctance to buy combed wheat reed cooperatively. He preferred to go to a dealer, 'who, he said, catered for the individual requirements of thatchers'. Some of the comments at that first meeting reflected anxiety as to whether there would be enough house thatching work for survival and there was some reference to the very recent agricultural past from which they were departing. Mr Edwards:

> also did not agree with the suggestion that thatchers should take to hedging and ditching as a bad weather occupation, contending that weather which was good enough for hedging and ditching was good enough for thatching.
> Mr Mitchell stressed the difficulty of getting thatching done and felt that the county should set an example by sticking to thatching on its own thacthin [sic] buildings and not substituting galvanised iron. (ibid)

The resolution to form the Somerset Master Thatchers' Association was put and voted on: 'there were 18 for and none against, and the resolution was therefore carried' (ibid). An executive committee of eight was voted in. Mr Sheppard outlined the Ministry of Labour's Vocational Training Scheme and the chairman closed the meeting with some remarks stressing the need for comradeship among craftsmen.

Items at the early meetings in Somerset included efforts to get hold of better quality tools, discussion of quality, pricing and contracts (the Bureau put forward a standard form of contract which thatchers could use), talks on insurance, social outings and frequent references (from the Bureau's representatives) to the desirability of a National Association.

This must have been a fairly typical beginning to one of the regional Master Thatchers' Associations.

By 1949 the newly-fledged Master Thatchers' Associations were strong enough to merit a national conference, the first ever national meeting of thatchers, held at the Bureau's headquarters in Wimbledon and attended by Cosmo Clark, the Director. Some of the themes debated would be familiar to thatchers today, as the minutes reveal. 'The effects of artificial manure on the quality of straw' was the first item on the agenda, followed by 'Types of wheat yielding suitable straw for different styles of thatching'. Opinions on this differed widely throughout the counties. The eastern counties preferred *Little Joss*, *Squarehead Master*, and sometimes *Yeoman*. The West Region favoured *Red Standard* or any red wheat, including *Steadfast*, with *Victor* regarded as a very good standby. The Midland Region favoured *Holdfast*, which was grown principally on light land. Wales favoured any red wheat. Debate on the quality and supply of thatchers tools, insurance and the desirability of forming a National Federation of Master Thatchers' Associations followed. The regional associations attending then each gave a brief resume of how their groups were progressing and the proceedings were wound up by Mr A J Barber, Secretary of the Master Saddlers and Leather Goods Federation, who had been invited to speak from experience on the formation of a trade federation and the way it functioned.

It is plain from published and oral sources that there was a wide variation in the concerns, success and effectiveness of the regional associations. There was, and is, no obligation to join the local Master Thatchers' Association, and some attracted a greater proportion of local thatchers than others. Some associations were happy to produce local specifications for thatch at an early stage, indicating minimum acceptable standards; other associations (eg Devon) were reluctant to do this. Some did indeed cooperate in efforts to drive down the price of materials, or to buy spars in bulk if they were in short supply locally, or to get special deals on tools and wire netting. Setting prices for work in any one region was always contentious because standard prices could be undercut by thatchers outside the Association.

Sid Pearce and other thatching members of his family did not join the Wiltshire and Hampshire Association, considering that the members were not the best craftsmen. Some extended thatching families in Devon took the same line towards the Devon and Cornwall association. Individual thatchers certainly saw benefits in association and recollect the early meetings as a remarkable and commendable revolution (Jack Dodson, pers comm). Beforehand, a thatcher would talk about the thatcher in the next village, but never to him. Afterwards, there was a level of cooperation, where a busy thatcher might sub-contract work to one who had less work, or suggest farms where ricks of good-looking thatching straw had been noted. Possessiveness about territory and skill before the Master Thatchers' Associations were set up is legendary (and perhaps exaggerated) among thatchers from all over the country, and there is no doubt that thatchers were secretive about their own craft skills. Chris Stanford from Cambridgeshire reports that his father would indeed hold the thatch upside down, or simply stop working if it was suspected that another thatcher was trying to learn the secrets of his trade by observing him (pers comm)

If the Master Thatchers' Associations were variable in quality and numbers in the 1940s, several of them more or less faded out in the late 1950s, and in 1959 it was reported that 'some had failed', although most were re-invented in the 1960s (Public Record Office 11). Today they are still variable in size, in strength and in the percentage of members, which appears to be less than half the known thatchers working in any area. All the regional associations have spent considerable time and effort discussing the value, make-up and effectiveness of the various national bodies that have been set up since the end of the war.

One of the most trying tasks that faces the Associations is dealing with customers' complaints about the quality of a member's work. In a small organisation where not all members necessarily agree on every aspect of thatching, this is a request to check out the work of a colleague, who may also be a competitor and accusations of bias are inevitable. Sometimes this can be resolved by asking a member of a neighbouring Master Thatchers' Association to carry out inspections. Accusations of bias are sometimes also made in the case of checking the work of applicants to the association.

In spite of these problems, unsurprising in small organisations, the regional associations have been crucially important to the industry. The fact that they have endured, whereas the first national associations had more trouble in surviving without a change of constitution, reflects the continued regional character of thatching. The problems in the supply of Norfolk reed have only indirectly impacted on Devon and Cornwall, for instance, counties where most water reed is obtained from foreign sources, and the details of the debate about the quality of combed wheat reed is only of theoretical interest to a thatcher who does not use it.

It is very doubtful that the Master Thatchers' Associations would have developed without the intervention of the Bureau. They have continued to have close links with the Bureau and its successors, and a representative from the relevant government agency usually attended meetings until recent times.

In 1994 there were thirteen Master Thatchers' Associations in all. Wiltshire was something of an oddity, since an older Master Thatchers' Association also covered Wiltshire and Hampshire. Qualification for entry to a regional Master Thatchers' Association involved inspection of an applicant's work by members to ensure that all members of an association had a standard of work acceptable to their peers. Their membership in 1994 is shown in Table 7.

There are pressures to join an association today, particularly for thatchers who do not spring from a thatching dynasty, or who are new to an area. While the fact that a man's or woman's father and grandfather may have been good thatchers is absolutely no guarantee *per se* of quality of craftsmanship, house owners may be impressed by the

Table 7 Membership of Master Thatchers' Associations, 1994.

Master Thatchers' Associations	number of members in 1994
Dorset & Cornwall	26
Dorset	23
East Anglia	25
East Midlands	11
Gloucestershire, Herefordshire, Warwickshire and Worcestershire	12
Kent, Surrey and Sussex	10
North of England and Scotland	4
Northamptonshire	9
Oxfordshire, Buckinghamshire and Berkshire	9
Rutland and Leicestershire	11
Somerset	10
Wiltshire	8
Wiltshire and Hampshire	11

genealogy. 'New' thatchers may be asked by clients if they are a 'master thatcher' (Gordon Elston, pers comm) and this may encourage them to join either the local Master Thatchers' Association or the National Society (see below). In fact the phrase 'Master Thatcher' has been much devalued, since anyone can call themselves a master thatcher and, outside the local Master Thatchers' Association or the National Society, which inspect an applicant's work by members to ensure standards (although the latter do so by convention only, not as part of their constitution), it is meaningless as an index of quality of craftsmanship or even length of career. Including the 'master' in a trade name does not mean that the thatcher working on the client's roof is a member of any trade organisation, or that the proprietor of the firm is a member of a trade organisation. A trade name including the word 'master' can easily be confused with local trade organisations, another possible pitfall for a client.

THE EVOLUTION OF NATIONAL TRADE ORGANISATIONS

From the outset of the Regional Master Thatchers' Associations, the Bureau was keen to encourage a national body for the industry. The first constitutions of the Master Thatchers' Associations had the prospect of a national body written into them.

> In the event of a National meeting or Conference two members of the Committee shall be elected to attend. Travelling expenses 3rd class and reasonable out of pocket expenses will be paid by the Association. (Somerset Master Thatchers' Association Archive)

A recognition of the status of thatching as a fully-fledged industry depended on a national voice, the development of a such a voice and some sort of administrative centre. In the absence of this, thatching was bound to depend on an 'outside' centre, like the Bureau or later, the Council for Small Industries in Rural Areas. The Bureau outlined the advantages of a national body in 1948.

> By this means thatchers for the first time could speak with a united voice on matters affecting their own craft; they will be able to increase their efficiency and down the shoddy work of pirating straw hangers. (Rural Industries Bureau 1948, 30)

The early minutes of the Regional Associations have frequent references (made by the Bureau's representatives at the meetings) to the desirability of a national body, but these met with little enthusiasm from members. It became a waiting game, with the Bureau hoping that the Regional Associations would gain in strength until a national body became inevitable.

In the end, it was not a thatcher, but one of the Rural Industries Organisers, Graham Castle of Northamptonshire, who initiated the National Federation of Master Thatchers in 1961–2 and became its Acting Honorary Secretary. This was reported in *The Guardian*, 22 March 1962, as being based on Warwickshire, Northamptonshire and Oxfordshire. One of the principal local motives for federation at the time was problems in the supply of materials, as well as anxiety about the poor public image of thatching. The supply of materials was a straw problem. The price for combed wheat reed had risen to £51 per ton. The Federation claimed that it improved the supplies of long straw by having it 'brought into the Midlands' (presumably from other areas) and that reed combers had been obtained in Gloucestershire (1961) and Northamptonshire (1962).

The National Federation was intended to be an association of local Master Thatchers' Associations but it never attracted all of them. A body representative of the whole trade was slow to evolve. By 1963 the nature of federation and how local Master Thatchers' Associations would belong was still being debated (Somerset Master Thatchers' Association Archives, minutes of a meeting 9/6/63). It was not until the late 1960s that a National Association of Master Thatchers came into being, following a conference in London in 1967.

> The main object of the conference was to form a National Federation of Master Thatchers' Associations and this was well under way. (Somerset Master Thatchers' Association Archives, minutes of a meeting 7/10/67)

THE NATIONAL SOCIETY OF MASTER THATCHERS' ASSOCIATIONS

The Development Commission offered the proposed National Society a grant towards setting up and the Council for Small Industries in Rural Areas helped to draft the constitution. There was an executive council formed of representatives of each constituent Master Thatchers' Association. Ten Master Thatchers' Associations applied for membership, giving an overall membership of 114 thatchers with a fee of £2 2s per member.

The Development Commission's grant was dependent on the Society cooperating with the Council for Small Industries in Rural Areas, and on efforts by the Society to make itself self-supporting in the long run, with a view to handing over to the Society the Council for Small Industries in Rural Areas's experimental Thatching Apprenticeship Training Scheme.

The National Society of Master Thatchers Associations took some time to evolve as a body with a settled constitution. Although it was first formed in 1967, amendments to the constitution were still being discussed in 1969. These related to the difficulty of categorising thatching, a re-emergence of the problem that the Bureau had faced in the 1940s. This time it was not a question of whether it was an industry or not, but what sort of industry it was: agriculture, or construction? The Department of Employment and Productivity was keen on the establishment of a national trade voice in order to simplify the training of thatchers, which it had decided should come under the Construction Industry Training Board, although previously rates for trainee thatchers had been in line with the Agricultural Board. The wording of the draft constitution of the National Society included reference to the Construction Industry Training Board as the body under which training thatchers would operate. This would have brought thatchers under the Selective Employment Payment Act of 1906, meaning that thatchers who employed others could not claim tax refunds for their employees.

The members of the National Society were unhappy about this and reluctant to accept the draft constitution. On philosophical grounds they did not want to be classed as part of the construction industry. On financial grounds, the larger firms wanted the tax refunds that they would get for employees under the Agricultural and Horticultural Training Board, but not under the Construction Industries Training Board. The philosophical grounds were outlined in a letter to the chairman of the steering committee of the proposed society, Mr Hobson, of Oulton, Norwich. George Dray, a Devon thatcher, expressed what many thought in a letter to the executive committee.

> It seems to us that thatching of all kinds is to be treated as an activity within Agriculture and Horticulture; and work done in obtaining the raw materials would be as much part of the activity of thatching as the operation of putting straw on to the roof. (Public Record Office 18)

The Council for Small Industries in Rural Areas felt unable to dispute the Department of Employment and Productivity's definition of thatching as part of the construction industry and the grant offered to the embryonic society went on hold while the executive committee argued about whether the wording in the constitution could be altered to 'the appropriate Training Board' rather than the 'Construction Industries Training Board'. In 1969 Hobson was still struggling to persuade the executive committee to accept the wording in the draft constitution, hoping that the question of what Training Board thatching was to come under could be sorted out later.

In March 1971 the National Society of Master Thatchers' Associations held a residential weekend rally in Great Yarmouth and recorded that eleven of the twelve Master Thatchers' Associations had joined the Society. It had been hoped that the Agricultural and Horticultural Training Board would assume financial responsibility for the training courses at Knuston Hall, but this had not happened, and the Council for Small Industries in Rural Areas had stepped in and funded the courses themselves. Twenty-four apprentices, journeymen and mastermen had attended Knuston Hall 'for instruction in Norfolk reed and sedge work'.

By 1976 some members felt that the National Society was not really going anywhere. Jeff King, a thatching officer with the Council for Small Industries in Rural Areas, retired as Honorary Secretary. Membership dropped and attendance at annual conferences, which had sometimes been in excess of 100, declined. The society decided to alter the nature of membership in 1977, which was to be based on individual application, rather than on representation of the regional Master Thatchers' Associations. Its title changed to the National Society of Master Thatchers.

THE NATIONAL SOCIETY OF MASTER THATCHERS

By 1978 membership of the National Society stood at around 30 (with nine from Somerset). The Somerset Master Thatchers' Association resolved to stop paying '£10 to a cause that did nothing'. The National Society of Master Thatchers conference in 1978 was attended by 25 'of which eight were wives' (Somerset Master Thatchers' Association Archives, minutes of meetings 18/1/78 and 6/4/79).

The National Society of Master Thatchers is now based on personal membership. In 1994 it had 57 members, according to a list provided by the chairman, but this was said to need revision. Membership included individuals who belonged to the largest thatching firms in the country. Membership, by convention, required a similar inspection of work as required by the Regional Master Thatchers' Associations, although the constitution did not set this out, requiring only that an applicant be elected as a member by the executive council. A Master Thatcher was defined as a person who had served 'an accepted period of training and is now recognised by this Society as a master of their craft' (Membership and Affiliation in the Constitution of the National Society of Master Thatchers, 1 January 1991).

Since 1989 the National Society has run well-attended one-day seminars for conservation officers and invited guests. This has been a welcome opportunity to meet and discuss issues informally with thatchers as well as disseminating relevant information. These seminars have been one aspect of the seriousness with which the Society has taken public relations and, under the chairmanship of Christopher White, the National Society has been active in commenting on conservation policy documents and making its voice heard in the press.

THE NATIONAL COUNCIL OF MASTER THATCHERS' ASSOCIATIONS

The National Council of Master Thatchers' Associations was established in 1987 at a time when some thatchers felt that the National Society was becoming ineffective due to lack of support. Ironically, the establishment of the National Council of Master Thatchers' Associations, which was encouraged by the Council for Small Industries in Rural Areas, prompted something of a revival in the National Society of Master Thatchers. In 1994 the National Council of Master Thatchers' Associations was made up of eleven representatives, one each from eleven of the existing affiliated Master Thatchers' Associations. The nature of membership via affiliation was a revival of the early National Society, before its constitution was changed and membership became personal. All members of the National Council and all the thatchers they represented had reached a standard of skill that was checked according to the constitutions of the regional associations. The Northamptonshire Master Thatchers' Association was not affiliated, neither was the relatively recent Wiltshire Master Thatchers' Association.

The National Council of Master Thatchers' Associations produces detailed literature for owners, including some interesting references to thatch performance and advice on essential roof construction elements for thatch, suitable for architects designing a new thatched roof. Its published literature recognises a *locus* for building conservation in thatching.

The National Society of Master Thatchers represented about 7% of the trade in 1994. The National Council of Master Thatchers' Associations represented about 19% through the system of affiliation. The existence of two societies of thatchers at a national level is an awkward arrangement for any institution or individual seeking to research, discuss or develop aspects of the trade. Outsiders who are unaware of two bodies representing thatchers risk offending one by failing to contact both. There can be no doubt that the perceived doubling of effort also makes it more difficult for the trade to pursue research funding on questions important to its development or to mastermind effective public relations. A single, strong body, working on behalf of the whole trade, is something that the Rural Development Commission and many thatchers would be relieved to see.

An outside body, looking for a centre to the trade in 1994, would have found that the bulk of thatchers belonged to no trade organisation and had no formal contact with a trade body. These individuals and firms may have been the best tradesmen in a particular area, or the worst (including individuals who may have failed to have been accepted as members of either the local Master Thatchers Association or The National Society of Master Thatchers), or may be a mixture of both as well as some in between. The small number of thatchers in any trade body parallels the situation in The Netherlands. Of about 380 working thatchers there, less than half belong to the association of professional reed and straw thatchers, Vakfederatie Riet-en Strodekkers.

6 Training, 1940s–94

TRAINING BEFORE 1940

Before the 1940s the training of thatchers was undertaken by an experienced thatcher teaching a novice (often a son or nephew) on actual thatching jobs. In thatching families, fathers were often keen to see their sons follow in their footsteps and the children of thatchers often found themselves spar-making or twisting straw ropes in the evenings after school and in the school holidays. Before 1940 training in straw thatching usually included not only work on dwellings, farm buildings, ricks and clamps but was likely to be mixed with the acquisition of other agricultural skills: hedging, ditching or sheep-shearing, for instance. The length of time this method of training took, and its success varied. Nevertheless, most thatchers who were trained by relatives before the war reckon that six or seven years were necessary before a level of competence had been reached to allow the trainee to work independently as a masterman. Much depended on the thatching and teaching skills of the experienced thatcher and the aptitude and enthusiasm of the trainee.

More than one thatcher working today claims to have learnt how to put the material on the roof only when father was asleep after a lunchtime session in the pub, which gave a rare opportunity for actually getting up the ladder and on to the roof. Sometimes sons had no real interest, with results that were miserable for both pupil and teacher:

> Pulling handfuls of wet straw from a heap, forming a neat straight row from right to left, may appear elementary, but there is much more to it than that. Your head is often near the ground in the quartering part of the job. I was often assailed by sickness when, with lowered head, I got a smell of wet straw, which made me retch ... When dizziness forced me to give in, Father always said I was bilious. 'Sit down a bit. Bin eatin' b★★★★ sweets. Ya'll soon git over it.' Glad to rest in the dry straw for a spell, I would sense him hurrying into work in an effort to fill the gap I had caused. He was more anxious about that than over my personal suffering, and before long he would ask after my health, more in anger than in sorrow. 'Don't ya feel better now?' he would command, while my little bit of world, with the nasty tangy bed of wet straw as a blurry centre-piece, whirled round before my eyes. (Brown 1978, 77)

Trainees, starting young and usually living at home, accepted the demands of a tough system. One thatcher, who worked on an estate in Dorset just before the war, recollects several years assisting his father with both house and rick thatching before being allowed to work by himself on a barn. The barn was chosen for being 'out of sight', so embarrassing mistakes were not visible to anyone else. His father visited once a day and insisted on the removal of all the thatch, which was not up to his high standards, twice, before he was satisfied (Alan Fooks, pers comm). In the 1990s there was no time for such extended forms of teaching, but many thatchers would still claim that four years training and perhaps another two actually thatching are needed before a real level of competence is achieved, and that the learning process never really stops, since each new roof presents new problems. This is not to say that the odd individual with exceptional aptitude (or, occasionally, without) did not 'take up' thatching with very little instruction. There are records of individuals helping out a thatcher, or even 'having a go' at their own roof and then setting up in business.

In common with all trade apprenticeships, whether formal or informal, changes in education and expectation after 1945, as well as changes within the trade itself, began to undercut the viability of the old system of training. On his retirement in 1949, Arthur Farman of Farman Brothers, a water reed thatching firm, reflected:

> You can't learn thatching in six months, or in six years. The work is rather too rough for boys today. They want to work with their gloves on. (Farman 1949)

THE INFLUENCE OF GOVERNMENT ORGANISATIONS ON TRAINING SINCE 1940

The 1937–8 survey of thatchers (Public Record Office 4, 60–62) by the Bureau turned up only 10 apprentice thatchers in straw. The pressing need for young blood in the trade was dealt with in part by the 1940s Vocational Training Scheme (discussed in Chapter 4 above), but it should be remembered that this method of supplying an 'unknown' to an experienced thatcher applied to a small number of trainee thatchers; the rest were still selected and instructed by experienced men in the old way.

After 1949, when the first Bureau thatching instructor was appointed, on-site instruction was provided to experienced thatchers, at their request. This was a process requiring considerable tact on the part of the instructors, and a level of humility from the thatchers themselves. To an outsider, it is extraordinary that such a system worked at all, given craft pride and suspicion of government officials. The critical issue was the practical thatching experience of the officers as fellow craftsmen, and their diplomacy. The system was helped by the Bureau's encouragement of combed wheat reed into long straw areas, since thatchers could ask, or be prompted to ask, for cross-training in combed wheat reed, and the instructors would then have an opportunity to correct 'faults' of technique.

The instructors carried out an astonishing number of site visits, giving instruction, each year. In 1959–60, for example, 250 instructional visits were carried out by the three thatching instructors (Rural Industries Bureau 1960, 34). They often mixed their visits with advice on supplies of spars, or on straw. They might have noticed promising ricks of the latter *en route* to see the thatcher, and they sometimes carried supplies of tools for sale. Visits to individuals were mixed with more time-efficient visits to groups of thatchers that could be arranged via the regional associations of thatchers, as they developed.

By 1959–60 the Bureau was pressing experienced thatchers to take on indentured apprentices. Ten fully indentured apprentices were reported in 1960–61, and in the following year the Bureau's report refers to two candidates for an 'experimental' indenture scheme that they were piloting. Unfortunately no details of this scheme have come to light in the course of this research, beyond a reference to the fact that apprenticeship took five years (magazine cutting, *c* 1965, MERL).

Knuston Hall training courses

The Bureau started residential courses at Knuston Hall, near Irchester, Northamptonshire in 1965–6 in response to a request from a trainee thatcher from that county who pointed out that, unlike trainees in other trades, he was unable to go to night school to supplement the teaching of his masterman. One of the Rural Industries Organisers, Graham Castle, who was also a prime mover in the first attempt at federation of Master Thatchers' Associations (later out-distanced by the National Society of Master Thatchers' Associations), set about finding a location for courses and found Knuston Hall.

Special courses in Norfolk reed thatching in the first year included courses for apprentices. The following year, 1966–7, seven residential courses, four for apprentices and three for mastermen, were held, all on instruction in Norfolk reed. These courses were developed over the next few years, using full-size model roofs on which practical exercises were carried out.

There were a number of reasons why Norfolk reed thatching was the only technique taught at Knuston Hall until 1971. Fred Cooper, the thatching instructor, was a Norfolk reed thatcher. At the time the courses started he was of the opinion that water reed would eventually replace both the straw thatches. As Jeff King has pointed out (per comm), there was also a question of economy. Water reed is the most economical material to use for instructional purposes as it can be used repeatedly until it becomes too short with the tops broken off. Combed wheat reed can be used several times by trainees if it is not wetted too much and stored in a damp condition. Long straw is the most wasteful material as it can only be used successfully by one novice, although it is often used twice when the second user knows how to yealm.

Since 1968 certificates have been given for proficiency in the different materials, examined by one of the thatching instructors. The standard required to gain a certificate is somewhat higher than that to join a local Master Thatchers Association. A minimum of three completed roofs is inspected as well as one roof in progress. Any criticism is passed on to the thatcher and if he fails on the first or following attempts he will be told the faults and is able to apply again in one year's time.

In 1971 at the suggestion of Jeff King, a thatching instructor who had started teaching at Knuston Hall in 1970, training courses were extended to include both combed wheat reed and long straw. Trainees entering the scheme opted to study two chosen thatching techniques. The common pattern thereafter was for a trainee from the eastern counties to choose long straw and water reed while a Devon trainee, for example, would choose combed wheat reed and water reed. In the early days of the scheme some 1–1½ tons (1.016–1.524 metric tons) of straw for long straw thatching were purchased for instructional purposes while in 1989 4 tons (4.064 metric tons) were used. Quantities of combed wheat reed over the period averaged between 3 and 4 tons (3.048 and 4.064 metric tons) per year.

The courses at Knuston Hall were designed to be responsive to the needs of thatchers. The first practical courses were termed 'Open Courses' and these have continued to the present day. An individual, perhaps an experienced thatcher wanting to brush up on technique or cross-train in a different method for one or two weeks, could fit in with one of the scheduled courses on the training scheme. Should there be several requests for such a course, special arrangements could be made.

Training was not confined to putting material on the roof. Estimating courses, including general business management were available too. These courses were also held in different parts of the country, very often at the local pub, so that the whole industry had the opportunity to participate. Many thatchers, including some of the best craftsmen, lacked skills in pricing up and the business paperwork, including insurance and health and safety issues, became increasingly important to financial survival.

In 1973 a NETS (New Entrants Training Scheme) was introduced by the Council for Small Industries in Rural Areas, and was subsequently taken over by the Rural Development Commission in 1988. This led to a City and Guilds qualification. The courses consisted of twelve weeks block release tuition, undertaken over two years, with continuous assessment and a two hour written exam. Applicants needed a minimum of three months experience, preferably six months, with a masterman. At the end of the training period trainees were expected to be competent in at least two materials of their choice.

The courses covered eleven main elements, based on practical exercises carried out on model roofs, usually at Knuston Hall, beginning with the basics of eaves and coatwork and carrying through to complex features. In addition to the eleven elements, the syllabus included instruction on related aspects, such as safety, roof construction, crop husbandry and coppicing. There was no charge for the courses, and candidates' travel and accommodation expenses were met by the Rural Development Commission. A grant could be paid to the masterman in question, to compensate them for the time trainees spent attending courses. Between 1973 and 1988 approximately 90 students passed through Knuston Hall, and between 1988 and 1994 there were 102 students. The figures before 1973 are unknown (Rural Development Commission figures).

The various types of courses at Knuston Hall made the thatching instructors (as well as trainees), particularly from the 1970s onwards, keenly aware of the diversity of ways of thatching as they observed trainees and mastermen from different parts of the country using slightly different methods. There was a growing recognition that unless a level of tact and tolerance was exercised, Knuston could become a way of 'standardising' thatching technique. The instructors did their best to avoid imposing their own techniques on trainees, so long as they could see that alternative techniques would give an acceptable level of thatch performance on a roof. In some cases, where the work of trainees revealed that a masterman's method might jeopardise performance or speed, it was possible to amend this by teaching the trainee to be more efficient and hoping that the experience might be absorbed by the masterman back on site.

The Training Scheme for Novices by Thatching Advisory Services Limited

In 1982, a large firm, Thatching Advisory Services Limited, decided to run its own training courses for its pool of thatchers. These were trained at the firm's headquarters at Finchhampstead in Berkshire, using full-size model roofs, similar to those at Knuston Hall, for practical training exercises. Six month's intensive training of selected applicants was succeeded by a system in which the franchiser provided a support programme for the franchisees. This method of intensive training and follow-up support was one aspect of a business-like approach to thatching which generated considerable debate within the industry. It raised the question of the effectiveness of intensive training in relation to training on actual roofs under the supervision of a masterman and the optimum length of time for training.

TRAINING IN 1994

The Knuston Hall courses, up to 1994, were taught both by retired thatching instructors working as consultants to the Rural Development Commission, and by working thatchers. The Rural Development Commission saw the goal of the courses as producing thatchers capable of eventually carrying on a successful business (Peter Evans, pers comm). Although debate about the conservation issues frequently arose during teaching before 1994, and still does, at present the courses do not include any detail about, say, filling in listed building consent forms, even though conservation officers (eg Somerset County Council and Districts) often find it more effective to inform thatchers, rather than owners, of the interpretation of the legislation, and expect thatchers to take some responsibility for explaining the local situation to owners. The absence of a formal element on conservation at the Knuston Hall courses is a gap which could be filled to the benefit of owners, thatchers and conservation officers. An input into the courses from English Heritage, county and district conservation officers and/or members of the Institute of Historic Building Conservation (formerly the Association of Conservation Officers) could be a useful addition.

The National Vocational Qualification

In 1994 a scheme for a National Vocational Qualification in thatching was proceeding through the necessary bureaucracy. The nature and philosophy of the training has involved the Rural Development Commission, working in conjunction with an industry lead body, made up of representatives of the various Master Thatchers' Associations The constituent parts of the lead body were agreed upon by both the National Society of Master Thatchers and by the National Council of Master Thatchers' Associations. The industry lead body includes representatives of both the national organisations. This is a welcome example of co-operation between the two national trade bodies. The syllabus is designed to raise standards, while avoiding standardisation. Some aspects of the NVQ scheme have yet to be resolved, notably the cost of examination and the selection of examiners.

Part III

Changes outside the thatching industry

7 Changes in production and supply of thatching materials

STRAW: CHANGES IN FARMING FROM 1940

Machinery

See Chapter 3 for information on crop species changes. In 1938, Claude Culpin summarised the mechanization of harvesting in the first edition of his *Farm Machinery*.

> Fifty years ago the normal method of cutting the corn was with the scythe, and one man could then mow about an acre a day. With a binder and two teams of three horses each, 10 acres a day can now be cut and tied by one man; while if two men are equipped with a tractor and a combine harvester, they can cut and thrash up to 20 acres a day. This illustrates the imperative need for present-day farmers to appreciate the economic possibilities of mechanization. (Seddon 1989, 69)

Farmers' appreciation of the financial advantages of mechanized harvesting were given a fillip during the Second World War in the push for self-sufficiency in food. Remnants of pre-mechanized harvesting that had survived into the 1930s (hand-flailing was noted then in both Wales and parts of Devon) seem to have vanished after the war. Before the war thatchers complained that mechanized threshing and reed combing produced inferior material to the old ways of hand-flailing and hand-combing, but they adjusted to it. Thatchable straw could be produced via the reaper binder (Fig 21) and the threshing machine with or without the reed comber attachment (Fig 22). Hand-flailing is still carried out in Holland on rye straw, used for ridging in the Drenthe region (anon 1990, 50).

The next great stride in the development of harvesting machinery, the spread of the combine harvester, was one of the steps that took grain production on a completely different path from the production of thatching straw. The use of the combine harvester had a catastrophic effect on the supply of straw for thatching. These machines combined reaping and threshing in one operation (Fig 23) and left stubble no more than two feet (610 mm) high and often only a few inches (50 mm) tall. The threshing process damaged the threshed straw, making it completely unsuitable for thatching.

The combine harvester
The Ministry of Agriculture promoted combines during the war as part of the drive to efficiency in food production (Fig 23). The wartime machines were mostly pulled by tractors although self-propelled combines were available in the early 1940s. A small pamphlet published by the Ministry in 1941 was designed to answer anticipated questions. The only reference to thatching was to its agricultural use for covering ricks: 'Can combined straw be used for thatching?'. The answer given was 'Yes. It must be stressed however, that threshed straw doesn't need to be thatched as efficiently as a corn stack. Any rough thatching will do'. There was no advice on house thatching, which fell outside the Ministry's remit.

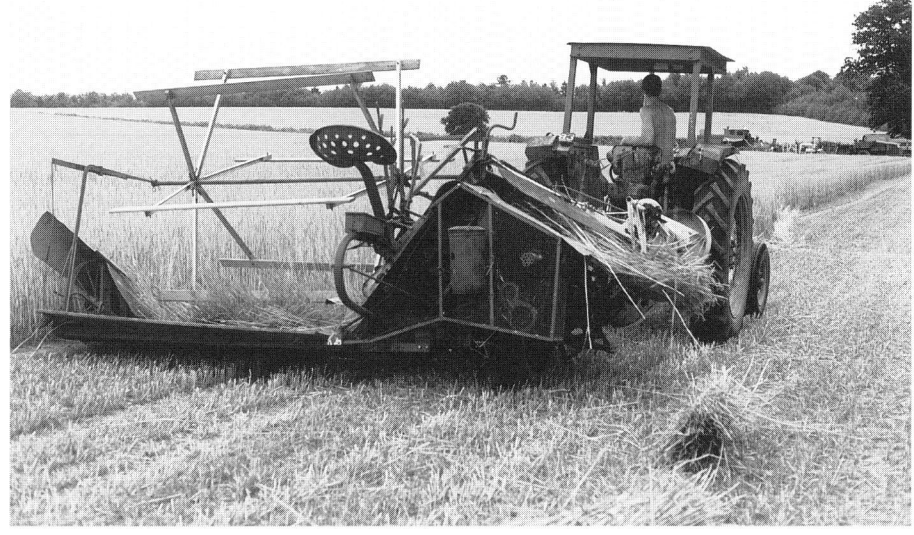

Figure 21 A reaper binder at work in 1991 (Keystone Historic Buildings Consultants).

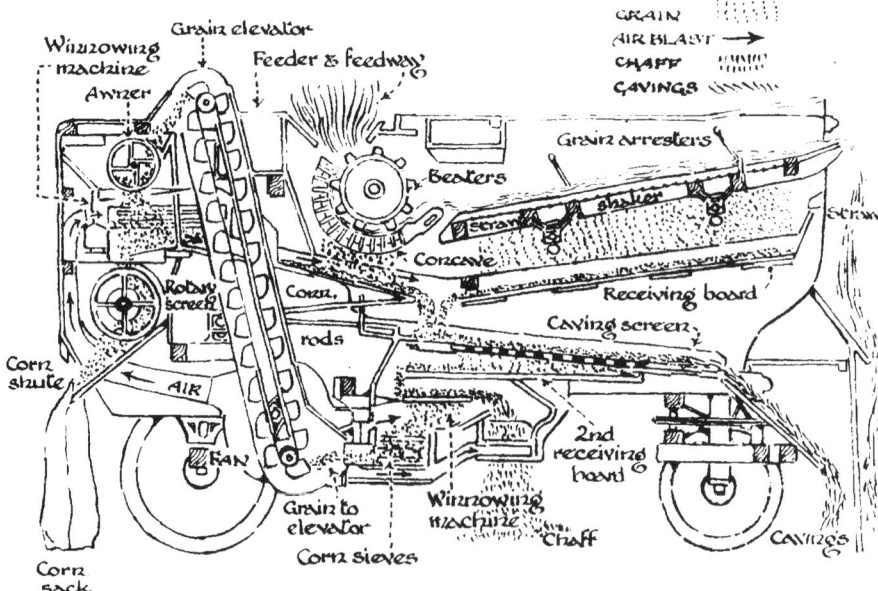

Figure 22 Section through a threshing machine (Ellacott 1981, 60).

With the intensification of farming during the war, there were 2,500 combines in Britain by 1944, more than 10,000 between 1948 and 1950, and more than 50,000 by 1960 (Brigden 1989). The combine had an uneven distribution. It was more popular in the north where there were labour shortages, and arrived latest, and sometimes not at all, in areas of small-scale pastoral or mixed farming, eg in Devon and Cornwall. It was not always easy to get such large machines down narrow lanes in some areas. The earliest machines were not only large, but so expensive that their cost could only be justified on farms with large acreages of cereal. Economies of scale meant that by the mid 1960s, 65,000 self binders were still being used on British farms (Collins 1969). However, the new harvesting technology had its impact on the machinery that was suitable both for processing wheat for grain and for producing thatching straw. Threshing machines more or less ceased production by 1950 (Duckham 1963). The reed-comber, the special attachment to the threshing machine for producing straw for the straw hat industry and also used for combed wheat reed, ceased production in 1952 (Gordon Glover, pers comm).

Along with developments in wheat-breeding, described by John Letts in Chapter 3, the spread of the combine ensured that the production of thatching straw gradually became confined to small, old-fashioned farms still using the reaper binder or evolved into a specialist business, where the priority for the producer was the production of straw, not grain. These farms were least likely to be in areas of intensive cereal production. Some thatchers began to take responsibility for producing their own straw, either by renting land and growing it themselves, or by being intimately involved with its husbandry and harvesting. This was pioneered in Suffolk by Frank Linnett, a well-respected thatcher who was not impressed by combed wheat reed used in Suffolk on former long straw roofs in the 1960s when supplies of good quality straw for long straw were particularly difficult to find. Linnett demonstrated that a thatcher producing his own material had the best chance of obtaining enough straw of the best quality available in any season. Treating the grain as the by-product he was protected from changes in the price of grain that increasingly tempted farmers to avoid the older and often lower-yielding varieties that produced the best thatching straw. Long straw is likely to be produced either by thatchers, or supervised by them or by specialist growers sensitive to the needs of the thatcher. This has certainly contributed to a general improvement in its quality in the last twenty years.

The development of the combine had other effects on thatching. In order to work efficiently the corn had to be cut at least a week after it was ready to cut with a binder, to improve harvestability and reduce moisture content in storage (John Letts, pers comm). This later time of cutting gradually became the norm. It is received wisdom that thatching straw is best cut green or 'gay' before the plant has died to maintain a level of strength in the straw that disappears as the plant ripens. The earlier time for cutting is now a departure from the norm which farmers are sometimes unwilling to make, particularly as early cutting is thought to reduce grain yield.

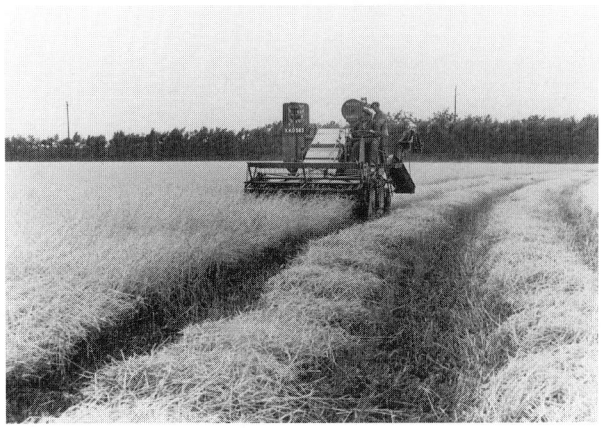

Figure 23 Combine harvester at work, 1956, Faversham, Kent. It was used for crops other than wheat, as seen here (barley) (Rural History Centre).

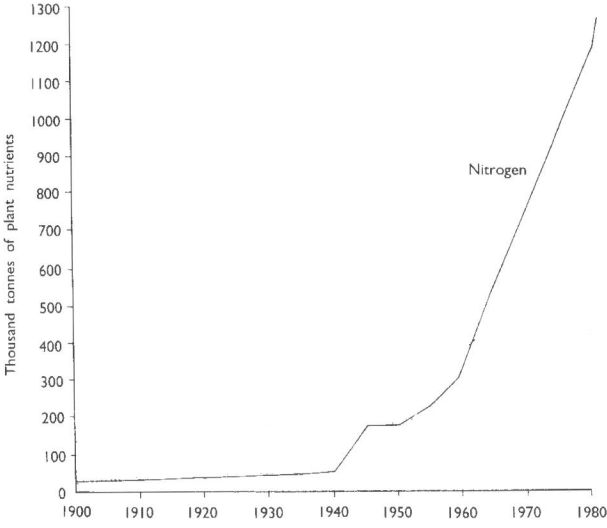

Figure 24 Nitrogen consumption in the UK, 1900–1980, after Grigg 1989, 72.

STRAW PRODUCTION IN THE 1990s

The business of producing straw today is a paradoxical mixture of ancient and modern techniques. It is hedged round with the legislation that affects all wheat-growing, described in detail in Chapter 3. Specialist straw producers today may be growing as much as 300 acres (121 ha) or as little as a single seven acre (2.83 ha) field. Harvesting techniques range from the most up-to-date equipment, the header stripper which, by chance, can be used for the production of thatching straw, to late Victorian methods.

Many traditional thatchers are reluctant to use straw from header stripper harvesting. The machine strips the plant of grain while it is still standing in the field. The straw can then be cut afterwards and collected into the round rolled bales that are a familiar sight in large cereal fields. The straw produced is usually used for combed wheat reed (John Letts, pers comm) although long straw thatchers have used it, in some cases after running over it with a tractor several times, to achieve the softness characteristic of material passed through an old-fashioned threshing machine (A Lewis, pers comm). The principal drawback of the header stripper method for producing thatching straw is the late timing of harvest required to make the machinery effective.

At the other end of the scale of straw producers, a farmer or thatcher may drill a single field with an old variety of wheat, sometimes in rotation with pulses to fix nitrogen in the soil. The crop will only have limited nitrates added, perhaps none. This is in stark contrast to the pattern of increasing nitrogen use in cereal and other farming since 1940 (Fig 24). Herbicides and fungicides may be avoided. A self-binder will be used for harvesting, the straw being stooked (Fig 25) and then ricked.

In the 1950s rick-making was considered one of the most skilful agricultural occupations, with a variety of shapes and finishes that were regionally peculiar. The results of competitions for the best rick were published in the provincial papers. Ricks allow the 'green' grain on the thatching straw to ripen and dry out without damaging the strength of the thatching straw. Rick-making for thatching straw is probably rather less artful in 1994 than it was in the past: labour is expensive and practicality rather than beauty is the goal. Constructing a rick, however, still requires great care to avoid the structure falling down and achieve the right amount of ventilation (Fig 26). Ricks seen in fields today are a sure sign that thatching straw is the object of the harvest. As an image that can be traced back through countless eighteenth- and nineteenth-century English paintings, it is a peculiarly moving aspect of the straw-thatching trade. The sheaves are drawn out of the rick for threshing after a number of months, and either threshed through a conventional drum (for long straw), or using a reed comber attachment to a threshing drum for combed wheat reed.

Today thatchers and farmers may find themselves competing with museums at farm sales for threshing machines and reed combers. Threshing machines, encased in wooden boards and Heath Robinson in appearance, may be worked during the week, powered by a tractor, but on show at a steam rally at the weekend, perhaps powered by a traction engine and admired as heritage technology. Watching the traditional production of thatching straw today is an unforgettable experience. Not only is the threshing machine extraordinary in comparison with most contemporary agricultural equipment, but the processes involve an input of manpower that is Victorian in quantity. Traditional threshing is difficult to carry out with fewer than eight men (Fig 27).

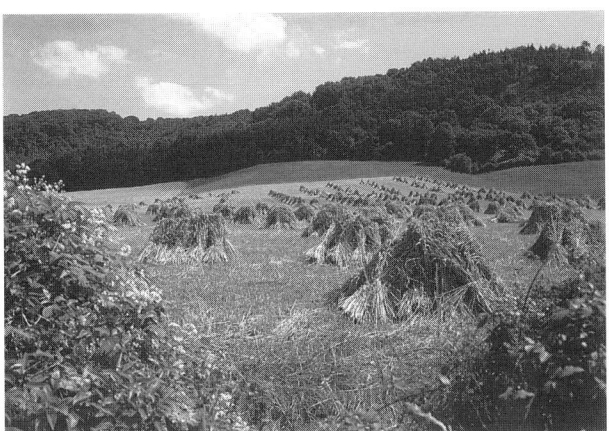

Figure 25 Stooks in a field used by the Wright family to grow thatching straw (Keystone Historic Buildings Consultants).

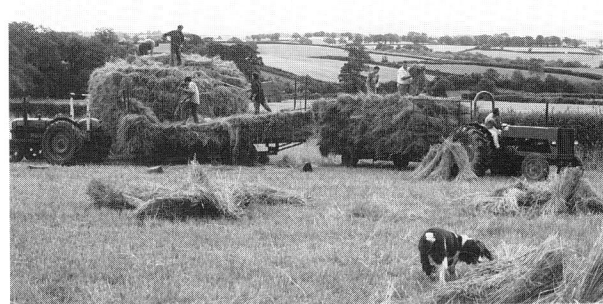

Figure 26 Rick-making in Devon, 1994 (Keystone Historic Buildings Consultants).

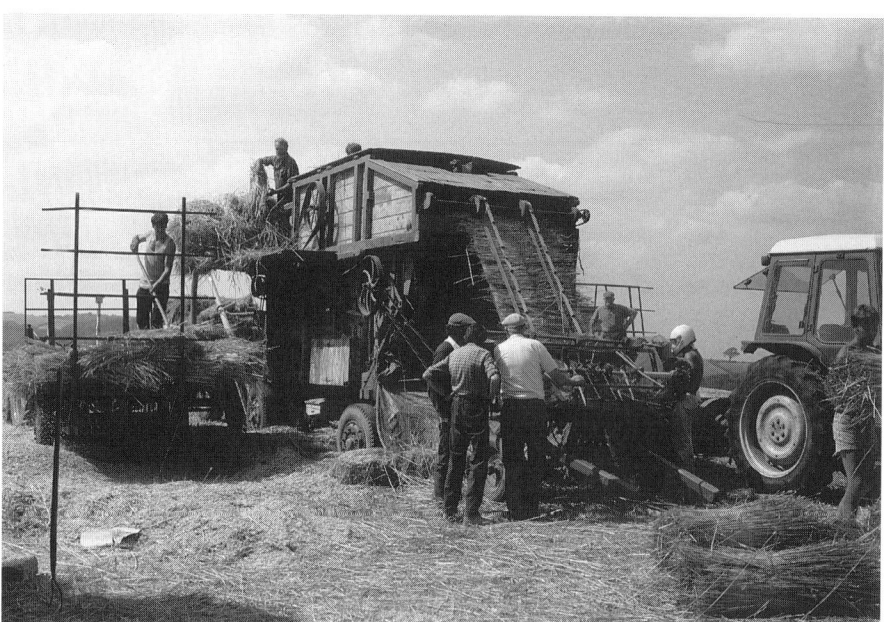

Figure 27 Tristan Johnson of Devon, thatcher, threshing with a team in Devon in 1994, using a reed comber attachment to produce straw for combed wheat reed thatching (Keystone Historic Buildings Consultants).

This is startling in an age when machines, rather than teams of men, do more or less everything on the farm. As one farm manager remarked 'These days my farm staff would use the mechanical digger to bury the farmyard cat' (Seddon 1989, 47).

Although harvesting with a reaper binder has to be carried out at only the right time, weather permitting, threshing can be carried out over a longer period. It is surprising that enterprising landowners and/or thatchers have not exploited the tourist potential of a scene that resonates with literary and visual references, even though it is an event that has to be timed according to the weather.

There must be some anxiety about how long the old-fashioned threshing machines and combers, many dating from the 1920s and 1930s, can survive the annual battering they take, often under far more pressure for speed than their inventors ever imagined. However, their technology, miraculous though it is, is said to be simple to repair and parts are simple to replace, provided the right carpenter or metal-worker (usually elderly) can be found. The numbers of threshers and reed-combers are finite, though. It is believed that Murch, one of the three firms who produced reed-combers, manufactured between 1,500 and 2,000 in all. There are about seventeen in Devon and a handful elsewhere (Gordon Glover, pers comm).

Between the largest and smallest producer of thatching straw there are gradations of scale, sophistication and ingenuity, as one would expect in any farming. There is sometimes an assumption that organic farmers are ideally placed to produce thatching straw, being familiar with low-input husbandry. However organic farms often have a greater than average need for straw for their own agricultural purposes and neither the motive nor the capital to invest in machinery that is specific to thatching.

One thatching firm uses mechanical Italian flower cutters in place of the reaper binder because they are considered to do the job adequately and the brand new machines that can be obtained require less cherishing than old reaper binders. There have also been various attempts to make the traditional Victorian methods of threshing more efficient with a view to the production of straw. To date, however, the size of the market for thatching straw has not encouraged expensive research and development in connection with harvesting and threshing, in parallel with the comparative absence of research on suitable varieties noted in Chapter 3. The Rural Industries Bureau and its successors have done more in this direction than anyone else. The Rural Industries Bureau commissioned an all-metal reed comber as early as 1950 (Public Record Office 9) and made available drawings of an old reed comber to straw producers who might want to commission one to be built.

The quality of straw available in any season depends on the weather and brings a considerable element of risk for the grower, whether a thatcher or farmer. A very dry season following a wet spring, for instance, will produce shorter straw than a wetter season, since the plants do not develop a good enough root system to put on the optimum length for the dry spell. If it pours with rain at the optimum time for cutting and the crop is left too long in the field, the straw quality will not be so good. If there are severe storms the crop may be damaged. No amount of goodwill, science or technology can resolve the problems that arise from a poor harvest, or a series of poor harvests, an age-old problem of farming.

CHANGES IN WATER REED PRODUCTION, 1940–94

The supply of sufficient indigenous water reed for thatching has presented problems since before 1940. Small quantities of reed were imported into Norfolk via Holland in the 1930s to supplement local supplies. The war

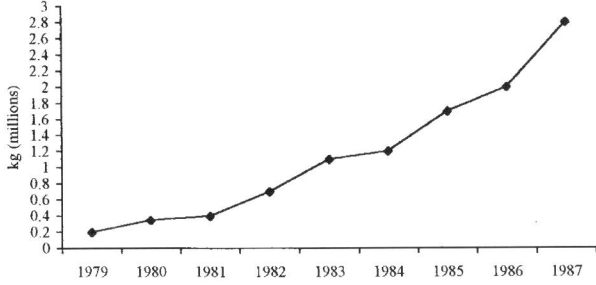

Figure 28 Volume of annual reed imports to the UK, 1979–87, after University of East Anglia 1991, 39.

intensified shortage. When imported paper was difficult to obtain, reed production was directed towards the paper-making industry and virtually dried up as a source of thatching material.

Demand for Norfolk reed increased after 1945 and it proved difficult to match demand with supply. The reasons were complex, and changed over time. To begin with, human reed-cutters were hard to find in the 1940s and 1950s: the job was recognised by everybody to be particularly unpleasant. In 1953 a survey by the Rural Industries Organiser for Norfolk gave the number of reed cutters as only five (Public Record Office 13). This may have been an underestimate, since 'unofficial' cutting was commonplace, but even if the actual number was double, the shortage of cutters was a real bottleneck in the supply of Norfolk reed.

Water reed is cut in winter, between December and April. Manual cutting, low on the stem, with a scythe was horribly hard work. 'It's a wicked job ... the cold fair cuts your legs off' (Cohen nd). As alternative sources of work, the sugar beet factories in East Anglia for example, began to provide year-round employment that was less arduous, the number of reed cutters dropped. In 1949, at the first national meeting of thatchers, it was agreed that reed cutters should be one of the allied trades allowed to join the regional Master Thatchers' Association, in an attempt to encourage interest.

Mechanization of reed-cutting was developed after the war, and helped to improve the supplies of reed. While new means of mechanical harvesting for cereals worked to the detriment of supplies of material for straw thatchers, technology applied to reed-cutting improved supplies for water reed thatchers. The Bureau's central role in developing and testing mechanical cutters in the 1950s has been referred to above and by 1963–4 50% of the Norfolk supply was cut by machinery.

There were also difficulties with the management of the reed beds. One of the management problems has been the history of a black market in reed. Water reed cut without the permission of (or payment to) the owners of the reed beds would be sold direct to thatchers. This has made the collection of accurate figures for reed production difficult. Some management problems are associated with what has been a perceived conflict of interest, between the conservation of wildlife and reed beds managed for commercial cutting. When reed beds came under the management of bodies principally concerned with the conservation of wildlife, eg the RSPB, the areas allowed for managing and cutting reed for thatching were often reduced. It has taken some time to confirm that commercial use for the production of thatching material has its own ecological advantages for wildlife, particularly birds and butterflies.

East Anglia supplies about one quarter of the water reed used in English thatching, the remainder comes from foreign sources. Since the shift to foreign sources began in earnest in 1950 there has been an increase in the use of imported reed and a massive increase in the last fifteen years (Fig 28). According to a University of East Anglia study, the value of reed imports into the United Kingdom in 1987 was about £1,000,000 (Environmental Appraisal Group 1991, 37), a nominal twelve-fold increase in volume and fourteen-fold increase in value since 1979 (op cit, 23), although precisely what percentage of this increase is absorbed by the thatching trade is uncertain.

At the time of writing Norfolk reed is actually cheaper than imported reed, the units in which it is sold being larger than the units used for imported reed. There are advantages to using native water weed, particularly for thatchers who are local to the reed beds. They have more control over the material by being able to choose it in the growing place and ensuring that they receive what they have actually chosen. Transport costs, however, and convenience for thatchers further afield are an issue, whereas the price of imported reed usually includes transport to the thatcher's chosen destination.

The movement to foreign supplies has made it difficult for East Anglia to catch up and compete with a sophisticated system of marketing and transporting water reed that serves a number of European countries, not just England. It may also relate to a desire on the part of some thatchers to abandon the business of a close involvement with the production of their supplies.

Foreign sources change with considerable speed. Turkey, which was a major source in 1989–94, especially in the south west, was not listed in the University of East Anglia study (op cit, 63), which looked at the rise in imports to 1987. Sources change for a number of reasons: the exchange rate, political upheaval or agents simply forging links with new countries of origin (Tristan Johnson, pers comm). Thatchers may find and prefer a particular source on the grounds that the water reed produced is closer in size and fineness to combed wheat reed, for instance, than another source. The shift from one country of origin to another makes it extremely difficult to keep a systematic check on longevity and quality of imported reed, since by the time figures might be available, a different source of supply is in fashion.

The real long-term concern for water reed supplies must be the tendency, all over the developed and developing world, to drain wetlands for more profitable agricultural purposes. Some Suffolk sources which supplied thatchers have been turned over to pasture since the war (J and B Death pers comm) and in the developing countries the speed of change is far faster and less likely to be constrained by what may be regarded as the luxury of environmental considerations. Wetland drainage could create real future difficulties both in the maintenance of quality of the water reed, and by pushing up prices.

8 Changes in demand for thatch

NEW OWNERS, NEW STATUS

Of all the influences contributing to changes in thatching since 1940, the changing nature of the owners of thatched rural houses has been critical. The difference between owners in 1940 and owners today is encapsulated in two quotations. The first is from Dorothy Hartley's book on rural crafts, *Made in England*, published in 1939.

> Thatching is ... essentially the work of country people for country people. (Hartley 1939, 50)

The second is from *Traditional Homes*, the 'essential' periodical for owners of old houses in the 1980s.

> [Thatch] is so beautiful to behold, it makes the home owner's heart leap for joy as he walks home from the station. (Hoppit 1985, 36)

Between these comments lies a period of change which has seen a completely new kind of owner, with a different culture, inhabiting and paying for the maintenance of thatched rural buildings.

Population change in rural areas 1940–94

As early as 1942, the *Report of the Committee on Land Utilisation in Rural Areas*, better known as the Scott Report, was noting 'drift from the country' by the rural population as a result of poor wages and poor housing. Improved housing for agricultural labourers was a main theme in the report, which included recommendations that the number of tied cottages be reduced and that farmworkers should be encouraged 'to have cottages built for their own occupation and with this end in view the subsidy provisions of the 1938 Housing Act should be more widely known' (Report of the Committee on Land Utilisation in Rural Areas [Scott 1942], Section 163).

The 'drift from the country' had a far greater impact over the next fifty years than the Scott report could ever have anticipated. As farm workers and farmers became fewer in number, previously tied cottages and, later, farmhouses themselves came onto the market and were bought by incomers who had a different set of cultural values from their predecessors.

Rural culture meant that farm workers who remained employed, familiar with the grind of cottages with limited water supplies, no electricity and poor sanitation, often preferred a modern building anyway and many went to the council houses built in the post-war programmes of social housing on village outskirts. There was some prejudice against thatch among working-class country people in the 1940s, on the basis that it was old-fashioned and insanitary, as expressed in 1947.

> Thatch is the material which townspeople sentimentalize over but do not use themselves. Most country people loathe thatch – 'rats' nest' they call it – and at the slightest chance they will rip it off and put galvanised iron or asbestos sheets to replace it. (Duncan 1947, 113)

Little reference to this attitude appears in published sources, which mostly reflect the very different views of the middle classes.

An exchange in the House of Commons in 1948 made reference to the problem of the survival of thatch after the war. Mr Skeffington-Lodge asked the Minister of Health if he was aware of the decay of thatching as a rural craft and asked what was being done to have repairs effected to thatched buildings.

> *Mr Bevan*: I am aware of the position, and I understand that my Right Hon. friend the Minister of Labour has arranged under the Government Vocational Training Scheme for training in thatching to be given by selected employers in an effort to maintain the supply of skilled workers.
> *Mr Skeffington-Lodge*: Is the Right Hon. gentleman aware that in the meantime the roofs of many charming thatched cottages, in such counties as Bedfordshire and Herefordshire, are being repaired with asbestos sheets, and that this arrangement is spoiling the attractiveness of many of the villages in which the cottages are situated?
> ...
> *Mrs Manning*: Will my Right Hon. friend look after these houses which have been repaired in such an ugly way? Since many beautiful cottages in Essex have completely disappeared, one agrees that it is better to repair the rest, but will the Minister see that they have their thatches put back as soon as possible?
> *Mr Bevan*: I have no prejudice against thatching. I lived for many years in a very agreeable thatched cottage. I will certainly look into the matter. (Hansard 1948, 1162–3)

It was particularly the middle-class incomers who were prejudiced in favour of thatch, campaigned to save it as a roofing material, and paid to save it.

The presence of incomers to rural areas was not immediate or extensive in the immediate post-war pe-

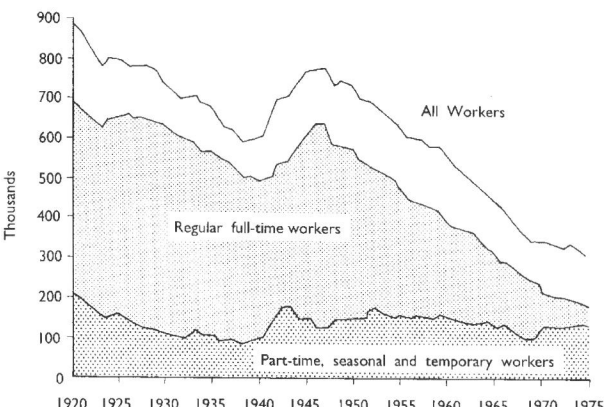

Figure 29 Changes in the number of agricultural workers, 1921–75, from Ministry of Agriculture Agricultural Statistics, England and Wales.

riod. Their arrival was the result of two factors. Firstly, increased distances became practicable for commuters to urban areas as road networks improved and, secondly, there was a significant and connected shift of labour markets away from the largest urban areas. The latter is termed 'counter urbanization' or 'population deconcentration' by historical geographers.

These processes have been at work since the 1960s, transforming the demographic map of England. In the 1950s, the overall population in the rural areas dropped by 0.06% for the decade, against a national average of a 5.0% increase (Champion 1989, 87). By the 1960s population in the rural areas had increased to 5.4% against a 5.3% increase in the national average. Between 1971 and 1981, with the national rate of population growth dropping to 0.6%, the picture in the rural areas was an 8.8% increase. No research has been found to correlate these figures and the numbers of people going out of farming, or shifting from tied cottages to council houses, but a relationship seems likely (Fig 29).

In some cases, thatch was the catalyst in the transference of ownership of rural properties to incomers. As agriculture changed and the thatcher became detached from the farm or the estate, the cost of straw was added to the thatching bill and this itself was likely to be detached from other work a thatcher might undertake for an agricultural client. As estates reduced in size after the war, the 'in-house' estate thatcher became a rarity, and the full cost of thatching, including the straw that had been free or waste, began to appear in the account books. In earlier days, some farmers had been bound by their leases to provide enough straw for the repair of their own buildings and sometimes enough, at a low fixed price, for the landowner to use on cottages.

As the cost of re-thatching tenanted properties rose and came to notice, it proved impossible for a farmer or landowner to offset the cost against rents from cottages on farms. If a local council required a property to be brought up to standard, it was sometimes cheaper and simpler for the landlord to knock it down than to re-thatch. Harold Wright, who thatched in Somerset, recollects receiving £120 for thatching a roof in 1950, although total re-thatching was unusual. With rents at 3s (15p) or 4s (20p) a week finding money for a large repair, let alone total re-thatching, could be difficult for an owner (Harold Wright, pers comm). The cost of re-thatching itself could be enough to push a local owner into putting that building on the market. An article of 1960 in *Farm and Country* makes precisely this point.

> Unfortunately, at Avonhead, we have a considerable number of thatched cottages, and each year the accounts show a disproportionate amount of expenditure on repairs to the thatch compared with the work on other cottages. (Hickish 1960, 13)

The figures to substantiate this are given. Given the estate's assessment of the average lifespan of long straw as 15 years 'allowing for the scarcity of good straw', the annual cost of re-thatching was judged to be £20. Some of the Avonhead cottages with a gross rateable value of £9 could only be let at £18, 'so that a landlord is bound to make an annual loss to meet the cost of thatching alone!' (op cit). The picture is the same today and one landowner in Dorset has a policy of selling thatched properties when the tenancy runs out, but retaining buildings with less costly roofs (Mr Pitt-Rivers, pers comm). Ironically, and providing another incentive to sell, the thatched properties are likely to fetch a higher price.

The sale of tenanted cottages to incomers was part and parcel of the processes of commuting and population deconcentration. These are only of interest here to the extent that they help to explain the very varied survival of traditional straw thatches. Villages and rural areas first drawn into commuting distances of large conurbations tended to lose their traditional straw thatches earlier than rural areas more remote from large towns and cities. A 1952 photograph of a house at Weston-on-Avon, Warwickshire, shows Norfolk reed thatch being put onto a house (presumably formerly thatched in long straw), which is described as symbolizing 'the change in occupation of many country cottages within easy reach of such industrial areas as Birmingham' (Rural Industries Bureau Collection, MERL, photograph referenced to the *Farmers' Weekly* Library).

As commuting distances expanded, the process of population deconcentration brought incomers actually working at some distance from the big cities. This was particularly true of the home counties and the Midlands, which were most profoundly affected by demographic changes in the 1960s and 1970s. The cyclical nature of population deconcentration appears to have pushed incomers into more distant rural areas in the 1980s. The 1980s exodus from London is close enough in time to be well-remembered. High property prices in London meant that it was possible to exchange a relatively small house there for a substantial rural farmhouse. A study of demographic change in North Devon, around the South Molton area (Bolton & Chalkley 1990), which was carried out to analyse the nature and origins of incomers, happened, in passing, to point out that they tended to occupy either the very oldest or the very newest buildings in the area. This reflected both a level of disposable income greater than local people possessed and a particular interest in old houses, if they could be afforded.

The new customer for thatch from the 1960s onwards had a real impact on the type of thatching used, and entered into contracts with the thatcher on a completely different footing from the farmer or estate owner who had paid for the vast bulk of thatching work before. The change in the nature of the customer was mentioned in the annual thatching reports of the Bureau, although given less prominence there than it probably deserved. The Bureau recognised that changing agricultural practice had disconnected the old web of informal, close, ties that linked the straw thatcher (whether of long straw or combed wheat reed) and farmer. Thatchers recollecting the 1930s and 1940s are able to give more detail about how the changes affected them personally, and remember the old days when farmers willingly took loads of straw (and a ladder), to houses that the thatcher arrived at on a bicycle, or later, an autocycle (Harold Wright, pers comm). Some farmers were willing to wait for payment for straw until the thatcher had been paid for his work on the roof. The arrival of the combine harvester and the Dutch barn put an end to a relationship of mutual favours mixed up with money and replaced it in a straightforward cash nexus.

The end of the close relationship between the thatcher and those who employed him changed as 'clients', increasingly, tended to be neither farmers nor their tenants. Thatchers found themselves for the first time working on houses which were not rented, but had been bought by incomers. It was not just that the thatchers no longer knew the occupants of the buildings and their circumstances personally, sometimes down to the rent they paid. Thatching the buildings of incomers became a matter of pure economic transaction, with profits to be made on the material, as well as labour, and it became possible to note, for instance, the first building that 'turned over £1,000' (Harold Wright, pers comm). The new importance of the cash nexus was spelled out in the script of a 1950 BBC Overseas Broadcast by a thatcher. The cost of a combed wheat reed roof on an average-size house was £130–£150. 'If, however, all materials are supplied by the farmer, it would not be likely to exceed £35' (anon 1950).

Incomers' enthusiasm for thatch was often described in comical terms. Houses have always been status symbols as well as places to live. A thatched roof demonstrated the reality of being in the rural idyll and was a badge of status as early as the 1930s.

> When they used the word 'thatch' a gleam came into their eyes – the same sort of gleam you see in a cat's eyes when she is watching a bowl of goldfish. Thatch they yearned for, and thatch they intended to have. (Street 1933, 753)

Increased prosperity and mobility in the 1960s converted a trickle of enthusiastic incomers into a flood. Articles in *Ideal Home* chart the hunt for rural properties in the 1960s and 1970s. In 1962 an article 'Running a Cottage to Earth' suggests Essex, Hertfordshire, Suffolk and 'the less accessible parts of Buckinghamshire' as 'good cottage-hunting country'. It illustrated a cottage thatched with long straw in Dorchester-on-Thames, Oxfordshire, costing £500, but pointed out that a similar property in Hampshire (with 5 acres) was much more expensive at £2,100 (anon 1962, 31). The reason for the price difference was that in Hampshire, as in West Sussex, larger estates and fewer small villages meant that small country cottages were more scarce and therefore more expensive. Cottage-hunters were advised to consult Hansard for Wednesday 12 July 1961, for details of road development schemes due to be completed, bringing a rural retreat closer to an urban centre.

By 1971 'Ideal Home Country Cottages' listed the most desirable locations as Berkshire, Surrey, Buckinghamshire, Sussex, Kent and Hampshire, followed by Norfolk and Suffolk. A third category, related to the pending M4 extension, included Wiltshire, north Somerset and Herefordshire. Readers were reminded of all that might need to be done.

> Common failings are the absence of a damp-proof course or floor concrete; dry rot; a roof that needs re-thatching. Thatchers are much in demand; journalists Mark and Sheila Jenner ... had to wait two years for the local craftsman to be free. The new roof [long straw] cost £240 ... and should last up to 35 years. (anon 1971)

As incomers became more numerous, they began to impact on the way the thatching was carried out. They did not necessarily know the old territorial etiquette that marked out one village or part of a parish as the province of one thatcher, not another, and unwittingly drew competitors into the territory by hearing of 'a good thatcher' from another incomer. The extension of thatching territory by recommendation from one owner to another had not been unknown before the war. Sid Pearce (pers comm) recollects travelling to Wales by train to thatch for a friend of the owner of the Wiltshire estate where he worked. These outlandish trips were rare for most thatchers working in combed wheat reed or long straw, although Farman Brothers, thatching in water reed, cropped up all over England including the Devon/Dorset border as early as the 1930s, as well as thatching in the United States and Germany. By the 1960s thatching far afield was more commonplace, and more prosperous incomers might call in Farmans, putting them up in a hotel or pub. Pubs themselves were often in the forefront of employing non-local thatchers, presumably because breweries, as corporate owners, were inclined to favour one or several firms of thatchers accustomed to travelling.

Preference for combed wheat reed and water reed

The question why incomers should have preferred combed wheat reed and water reed to long straw in the 1950s and 1960s is a complicated one. The Bureau's role in actively persuading incomers as well as thatchers to choose the material, or better still, water reed (if it could be obtained and afforded), was part of a broader change of taste. A preference for combed wheat reed over long straw, and water reed over both, was a matter of culture as thatch became a modern building material as well as a link with the past.

The supply of traditional straw thatching had been intimately related to local farming. Incomers were not part of this relationship. Not only was the working network with local people missing, but money had been set aside or was available for renovations. Expectations were that once money had been laid out on a scheme of renewal, it would not be spent again in a hurry. As the manager of an estate wrote in 1960:

> Many of the wealthier cottage-improving type of owner-occupiers accept the fact that thatching is expensive, and hope that it will not have to happen too often in their lifetime. (Hickish 1960, 136)

The material itself became a product, like any other in a marketplace, to be compared with alternatives. Durability, cost, convenience and appearance were the prime considerations. All the received wisdom in this context argued that water reed was the most durable thatching material, that combed wheat reed was second best and that long straw lasted less well than the other two thatches. This encouraged incomers to look for the first two types of thatch and avoid long straw. Water reed was the most expensive. As the work of Moir & Letts (Volume 5 of these Research Transactions) has shown, unlike the straw thatches, Norfolk reed had a longer history of being transported outside its immediate growing area. This added to the cost, as did middlemen between reed bed to roof. If it was more durable, however, the additional cost could be seen as justified, at any rate to an incomer who could afford it, especially when convenience came into the question.

Long straw is messy to fit compared with the other thatches. The straw is stacked in a heap, rather than in bundles and the process of yealming, preparing the material on the ground, tends to swamp a small garden with straw. The old system of getting maximum life out of a straw roof by a process of frequent patching, is repeatedly referred to by thatchers who were working before the war and into the 1950s. 'Re-thatching' as complete coatwork was rare relative to the amount of patching and partial re-thatching that was undertaken. Patching, however, was messy, inconvenient, and time-consuming, relative to no patching, especially for an incomer unfamiliar with the peculiarities of thatch. Precisely the same considerations could be seen at work in the change from using limewash, to proprietary paints, when they became easily available in rural areas. Older people in rural areas describe whitewashing their cottages, inside and out, annually, as an extension of the constant labour of housework. When proprietary paints became available, this was an aspect of domestic labour abandoned with some relief. Patching a straw roof fell into much the same category as annual limewashing, defined as inconvenient when alternatives were available, thrift was not the only consideration, and labour was no longer in-house.

Although water reed can be re-dressed to extend its lifespan, this was less common than the pattern of patching which traditionally extended the lifespan of a straw roof. As water reed did away with the inconvenience of regular patching, it had a substantial advantage in the eyes of the owner. Patching affected appearance, too, and was inconsistent with the neat, prosperous appearance of a roof, desired as a reflection of status. In 1954–5 it was said that 'a master thatcher's time is disposed approximately 3:1 between repair and maintenance/new thatch on both new and old properties' (Public Record Office 14). The pattern today would be a very different one. Incomers tended to expect roofs to look smart and even. Photographic evidence from all over the country shows that straw thatch in the 1920s and 1930s was patched, and often quite shabby in appearance. It is impossible to tell whether these roofs actually leaked, but in many cases they probably performed the function of keeping water out, which was regarded as the critical factor. Patching, or even investigating a roof to see whether patching is needed, is complicated by Health and Safety legislation on the use of long ladders that renders even the inspection of a roof a costly affair.

Appearance was another reason why combed wheat reed or water reed may have been preferred to long straw by incomers, although it is difficult to find documentation to substantiate this. It should be remembered here that the use of thatch for covering virtually anything that needed temporary weatherproofing on a farm was still in evidence in the 1950s and 1960s. The more temporary the covering, perhaps just a few months while a piece of equipment was not needed, the 'rougher' the appearance of the surface. Objections to the physical appearance of long straw, which is less neat and even than either combed wheat reed or water reed and uses the external fixing that was also used for very short-term thatches, was probably connected with ideas of status. To an incomer, a distinction in appearance between their house roof and a winter cover put over a grain elevator may have been an important one.

By the 1980s, there were signs of another revolution in the taste of some middle-class owners of rural buildings. Vernacular materials, details and surfaces increasingly began to be prized as authentic and irreplaceable in the same way that garden fashions veered back towards native or old-fashioned plants. This change is discernible in articles in *Traditional Homes* magazine, the bible of incomers with old houses in the 1980s, with technical advice offered by conservation experts. The 1980s saw a revival of maintenance-costly limewashes and lime-based products and a new emphasis on authentic surfaces with their more patchy uneven appearance. The revival of long straw, described variously as 'shaggy', 'hairy', 'poured on', 'ethnic', 'right for the building' was part of this revolution in taste. An interesting article in *Traditional Homes* in January 1985 picks up on this change. The typical thatch owner might be a commuter but could be expected to be knowledgeable about the choice of thatches and refers to long straw's 'own aesthetic charm'. A revival of thatch *nous* is also encouraged.

> Just because a roof looks a mess does not mean it is a total write-off. An expert thatcher could repair the ridge, dress the 'coat' and repair the bird holes, and have the roof looking (and functioning) as good as new. (Hoppit 1985, 36).

Other articles in the same magazine reflect an acceptance, at least among the cognoscenti of middle-class vernacular taste, for the input of the building conservation officer into the thatched roof debate.

> The Prudames decided to rethatch the whole cottage, together with the new extension. The local council wanted it to be done in long straw, and insisted that it should have a flush ridge without any frills or shapes. 'There is a conservation push to go back to what these cottages looked like in old photographs,' Paul said. 'Generally they look very plain. So some of the recent ridge patterns are a bit over the top' (McGlue nd, 5).

Thatch *nous*

The difference in culture between incomers and farmers as customers for and users of thatch was also a question of thatch *nous*, a difference repeatedly mentioned by older thatchers. Farmers and cottage tenants who had spent a lifetime under thatch were only too aware of the difference between an unsightly but insignificant birdhole or two, and a patch of degradation that was about to cause a leak.

Farmers, with long ladders to hand and, on small farms, used to thatching their own ricks, were quite capable of putting a patch on their own roof if the thatcher was not available. This DIY approach was not open to incomers in the same way and, in the 1950s and 1960s, encouraged a shift towards a types of thatch that required least maintenance.

Thatch *nous* also affected fire risk. It was second nature for individuals whose families had lived under thatch for generations, and who had seen their share of thatch fires, to take precautions to avoid fire risk. The open fire, often alight from one year to the next, might be put out for a week or two if a long hot summer suggested that the thatch was exceptionally dry. This may have been a misunderstanding of what is now known to be the comparative rarity of sparks from an open fire causing a thatch fire, but was a natural precaution. Wind direction would be considered if anything was being burnt on the farm close to the house.

Farmers might be careless with regard to wiring, as anyone who has regularly inspected the roof spaces of farmhouses will testify, but on the whole they had less of it than incomers and it was not uncommon, even into the 1980s, for roof spaces in farmhouses to be inaccessible anyway, before a mass of services were laid among the rafters, associated with a heavy level of plumbing and wiring to accommodate multiple bathrooms, central heating, electrical equipment etc. Farmers were used to cold houses and, certainly in the 1950 and 1960s, quite likely to live in the kitchen anyway, only heating the living room at Christmas. Woodburners, loved by incomers, sometimes with disastrous results, were just unnecessary expenditure for a farmer, who was more likely to be in bed by ten o'clock. Farmers were certainly unlikely to set fire to their houses by setting up a barbecue under the eaves, or pushing rocket sticks into the thatch on 5th November, occurrences noted by thatchers (Derek Wisbey, pers comm) called in to repair the damage.

THATCHERS AND OWNERS

As owners became more prosperous, so did thatchers, who by 1971 were beginning to look quite comfortably-off, to outsiders at any rate.

> Today the thatcher's new-found confidence, and the occasional Rover 3.5 litre car, springs from a rural industry which is far from down-at-heel or dying. If there is anything stirring in the thatch these days it's more likely to be money than rodents. (Prufrock 1971)

Whether or not thatchers had been badly off earlier on, is open to debate. Their wages before the Second World War had always been a cut above those of agricultural labourers, which was the obvious point of comparison in the 1940s. Harold Wright recollects that before the war, when more than half the thatching he and his father did was rick thatching, he was paid between 8s (40p) and 10s (50p) per rick, and could thatch one in a day. He could earn 9s (45p) per day making spars. Farm labourers were paid 6s (30p) a day. For him thatching wages provided a far better life than agricultural labouring. A good night out cost 2s 10d (18p): 4d (1p) for 10 Woodbine cigarettes, 6d (1½) for getting into a dance; 4 pints of beer at 6d (1½) a pint and nothing for going home, singing, on a bicycle (Harold Wright, pers comm). The author of an article in *Country Life* interviewed thatchers between 1930 and 1939 and discovered that many Norfolk reed thatchers 'worked only three or four days a week and one Berkshire straw thatcher said frankly that it was policy to keep two or three jobs going at once, because otherwise people thought he was asking too much for doing a roof' (Ward 1946, 529). This is probably highly exaggerated, and since the information about Norfolk reed thatchers was derived from an interview with a Devon thatcher, is probably no more than the usual tendency to put down craftsmen from a 'foreign' area.

There is no doubt, though, that from the 1960s incomers would pay handsomely for thatching, and thatchers' incomes could no longer be compared with cowmen's wages. It became possible to make a very decent living out of thatching and thatchers began to become more like their customers, as they had been like their farming customers in the 1940s.

Owners might initially have been fascinated by thatchers accustomed to a working agreement with the minimum of paperwork and a relaxed attitude to the week or month of arrival on site. Incomer businessmen, however, began to find it easier to deal with businessmen thatchers, who asked for contracts and interim payments, produced literature, or contacted them via a leaflet stuck through the letterbox perhaps offering a good value insurance package as well as a rethatch, and able to begin work next month. This was familiar ground and could seem more 'reputable', as well as easier to deal with.

Thatchers working today comment wryly on the difference between their way of life and that of the masterman who taught them. David Trezise was taught by a thatcher who rarely worked outside one parish and obtained his material locally. Not only has the area in which his former pupil works expanded vastly but, in the search for sources of good quality water reed, a passport and occasional trips to Turkey are required.

The relationship between owner and thatcher never followed quite the same path as that between an owner and any other building tradesman. Re-thatching was not carried out under the supervision of an architect. A thatcher was unlikely to share a site with another tradesman who might be critical, as tradesmen are, of other tradesmen's work. No-one interfered by arriving to inspect the standard of the work, or putting the thatcher in some hierarchy of 'experts' or regulations.

New owners were often wholly ignorant about thatch and entirely dependent on the advice of the thatcher. Many owners had tried their hand at DIY carpentry or brick-laying and might feel equal to a discussion with a carpenter or brick-layer about how best to carry out a job. There was little chance that an owner had 'had a go' at thatching, and the thatcher's word became law, the thatched roof his province and his alone. The owner might have enough knowledge of durability to insist on water reed and want a block-cut ornamental ridge (the latter said by thatchers often to be the only concern of an owner), but the rest of thatching was a mystery. If the thatcher said that wire netting was necessary to prevent bird damage, wire netting (commonly used on long straw since the Second World War, but also found on some combed wheat reed roofs) went on. If anyone was likely to resent the arrival of the conservation officer it was the thatcher, who could, if he wanted, claim (with justice) superior knowledge. Since conservation officers could only find out about the technicalities of thatching from thatchers, many of whom were reluctant to communicate with the local planning authority intruding on their territory, a divide began to open up in some quarters between conservation and thatchers, sometimes with the owner drawn into the argument.

Part IV

Conservation and training

9 Conservation issues

From the 1940s until the 1980s the principal conservation issue relating to thatch was whether or not thatched roofs would be replaced with other materials. Conservation efforts were directed at avoiding the loss of thatch per se, particularly on thatched houses that had been covered with corrugated iron or asbestos. This sometimes meant pulling in the opposite direction to other policies ie housing improvements. In 1957 a Rural Industries Organiser's report covering Bedfordshire, Hertfordshire and Middlesex noted that:

> The amount of thatch is rapidly declining as Councils condemn sub-standard property, a large proportion of which is thatch. (Public Record Office 8)

It was not only thatched houses that were at risk from government policy, but thatched farm buildings were excluded from improvement grants in the late 1950s (Public Record Office 8, 53rd meeting of the General Purposes and Finance Committee).

Planning authorities trying to save thatch in their areas were only too pleased when they prevented it being replaced with slate or tile. There was no great concern about what kind of thatch was used, but water reed was usually preferred if it was available. This view is still one that obtains in some areas. North Devon, part of a traditionally combed wheat reed county, was encouraging water reed as late as the 1980s on the basis that it was considered more durable than straw. Improvement grant clauses, still nominally current in some counties (eg Wiltshire) may specifically state that water reed thatch is preferred.

In spite of gloomy pronouncements, made since at least the 1930s, about the likelihood of thatch vanishing as a roofing material, there has been no sign of this since the 1950s. The status of thatch rose as the ownership of thatched rural buildings began to be transferred from country people to the urban middle classes. A measure of the value that is placed on thatched roofs is not only thatched extensions but the numbers of new buildings designed with thatch, a welcome example of extending the work available to the industry (Fig 30). The small but influen-

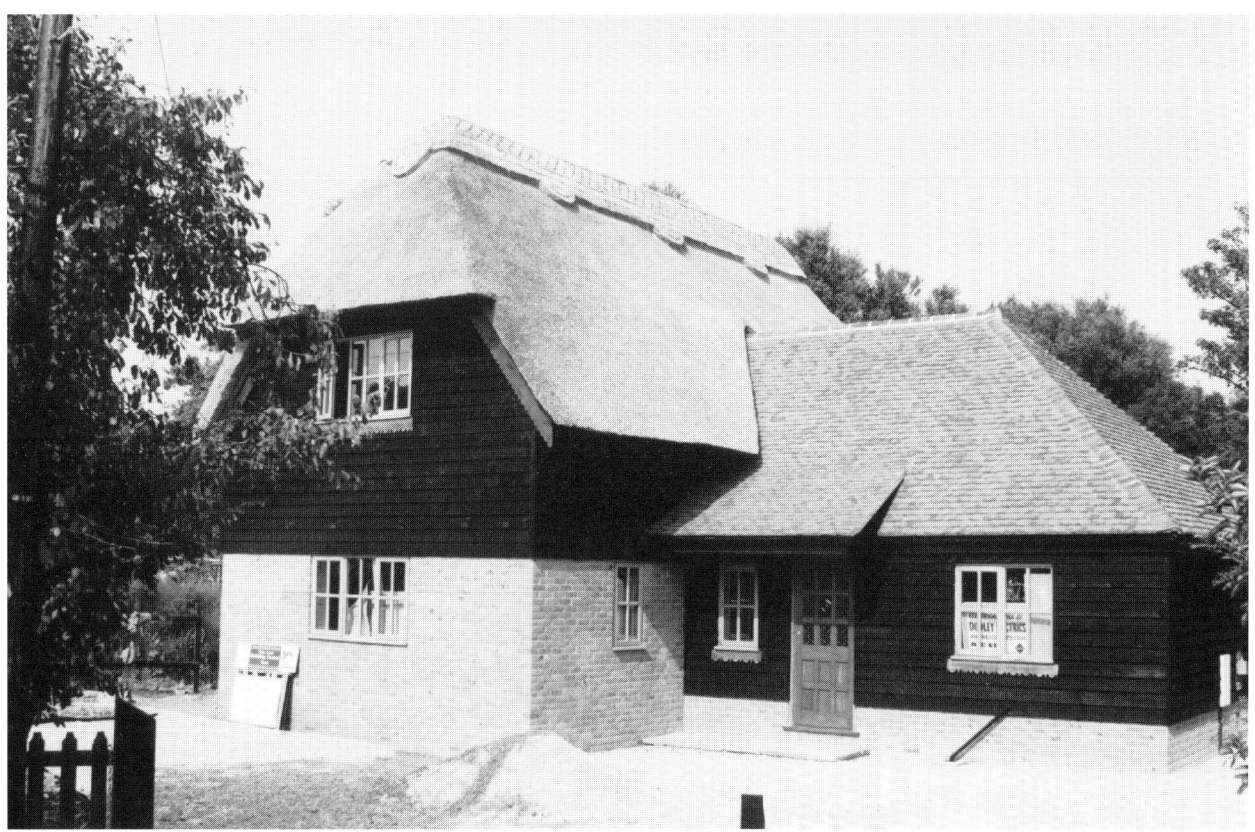

Figure 30 Thatch on a new house in Kent, 1994 (Keystone Historic Buildings Consultants).

tial strand of new houses designed for thatch before 1940 has strengthened in the past 20 years. The design of most of these houses makes some reference to the vernacular past, most convincingly in Dorset. The prospects for thatch on new houses has been improved by more realistic insurance premiums and specialist insurance companies. This reflects a better understanding of fire risk than was available before 1940.

Partly because thatch appears at present to be safe and unlikely to disappear altogether from the English scene, a number of far more refined conservation concerns than mere survival of thatch on old buildings has developed in the late 1980s. The reasons why the level of interest has risen so sharply are not difficult to supply. There has been increasing popular interest in all aspects of vernacular buildings since the 1940s, and a growing sense of their contribution to local distinctiveness, to the local landscape and local history.

A small number of academic historians in the late 1950s recognised the significance of traditional buildings as the material evidence of local and national history, filling gaps in the record of paper documentation. The Vernacular Architecture Group (The Secretary, Vernacular Architecture Group, 'Ashley', Willows Green, Great Leighs, Chelmsford CM3 1QD) was founded in 1954, dedicated to the recording, study and dissemination of information about traditional buildings, which gained respect as they were recorded and researched and their regional variations were better understood. The survival of medieval thatch, smoke-blackened from the open hearth fire, was first discussed and illustrated as material evidence for the evolution of medieval vernacular houses in a national journal by Alcock and Laithwaite in their seminal article in *Medieval Archaeology* in 1973, 'Medieval houses in Devon and their modernization'.

The potential breadth of vernacular buildings studies gave it a popular appeal, supported particularly from the 1970s by publications aimed at, or at least readable by a popular market. The nature of the subject encouraged the involvement of 'amateur' researchers and recorders, who were ideally placed to investigate their local areas and who became experts in the buildings of their own region as a matter of interest rather than paid work. Owners, as well as historians, came to place a higher value on the traditional buildings they lived in, reflected in their higher prices. By the 1980s the conservation of the structural elements of traditional buildings, employing techniques and materials that had fallen out of use in modern building practice, had become part of building conservation practice, with technical leaflets available to architects and owners alike from the Society for the Protection of Ancient Buildings (The Society for the Protection of Ancient Buildings, 32 Spital Square, London E1 6DY). A small and specialised corner of revivalist modern building practice developed, using traditional materials, whether oak-framing or versions of earth building, for repairs, extensions and new buildings. The practical knowledge gained from this revival fed, and continues to feed back into the study of traditional building materials.

On the government conservation front, the listed building resurvey of rural England in the 1980s for what

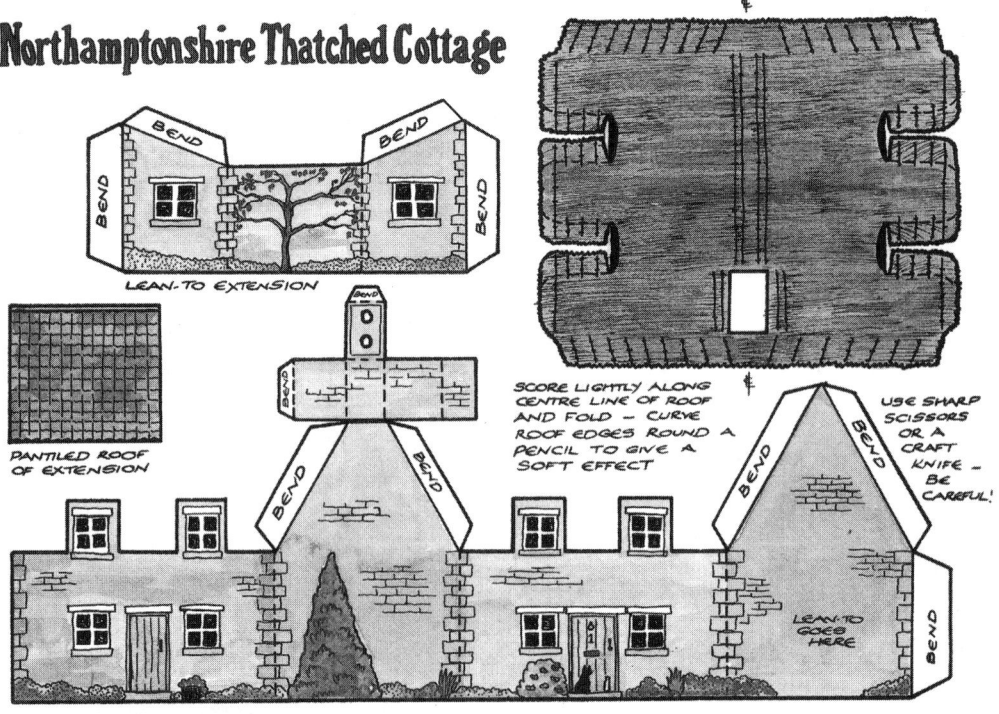

Figure 31 A postcard produced by the Northamptonshire Planning and Transportation Conservation Group. This was carefully designed as part of the movement to raise the profile of long straw in a county which had identified the rapid disappearance of this kind of thatch. It shows a typical east Northamptonshire cottage with a caption on the back explaining how the straw is yealmed and drawing attention to the simplicity of the ridge and the use of hazel liggers (Northamptonshire County Council).

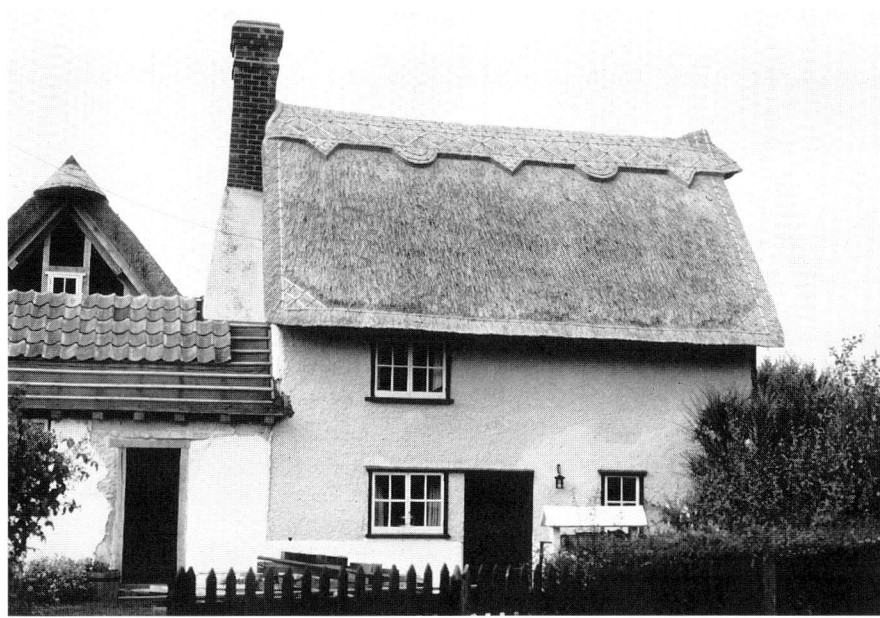

Figure 32 An ornamental ridge on a small vernacular house thatched in long straw in south Cambridgeshire, 1994 (Keystone Historic Buildings Consultants).

was then the Department of the Environment reflected the interest in vernacular buildings that had developed since the previous surveys. Before the 1980s, investigators covering rural parishes often identified only the parish church and 'big' house as being of special architectural or historic interest and worthy of protection. The 1980s resurvey vastly increased the numbers of vernacular buildings with statutory protection, and many of these were thatched. The figures for one parish in mid Devon, Stockleigh Pomeroy, were typical of some heavily thatched regions where pre 1700 houses were discovered to have survived in numbers. Before the resurvey of the 1980s there was one building with statutory protection, listed in 1965; the parish church. After the resurvey 23 buildings were listed, nine with thatched roofs plus three identified as having been formerly thatched. In a national context the resurvey ensured that many rural thatched houses were unlikely to be re-roofed with other materials, including those identified as having been thatched in the past. The resurvey provided assured work for thatchers on buildings with statutory protection, but it had other consequences too.

The statutory listing process brought thatching to the attention of conservation officers and planning committees. While the survival of thatch, especially smoke-blackened thatch, contributed to the interest of buildings that were given statutory protection, they were protected for reasons that included more than thatch alone, for the survival of pre 1700 fabric, for example. Once given statutory protection, the whole building is protected and this meant that conservation officers were bound to take an interest in any alterations that a change of thatch might bring to the character of a listed building.

Unfortunately, few, if any, of the field workers carrying out the initial assessment of buildings for listing in the 1980s gave any detail about the type of thatch that was on the roof when the building was inspected. This made it difficult for conservation officers to know whether re-thatching since listing had respected the type of thatch and its details at the time of listing. It took some time for conservation officers to become aware that the rate of replacement of long straw by combed wheat or water reed, in some regions, was at a pace that would obliterate the local distinctiveness of the thatching on their patch in a matter of years, rather than decades. Northamptonshire County Council was in the forefront of producing figures that showed how swiftly long straw in the county might disappear altogether. A survey of the Wellingborough area in support of a proposed County Council thatching policy in the late 1980s showed that almost 90% of roofs over five years old were of long straw whereas only 6% of roofs thatched since then were of this material (Fig 31).

Conservation officers became concerned about the arrival of heavily-patterned ridges in areas where photographic evidence indicated that vernacular houses before the Second World War had flush or plain block-cut ridges. In some cases the fashion for showy ridges could dominate the character of small or modest vernacular buildings (Fig 32). The frequency of grant aid requests on the same buildings also alerted conservation officers to the worrying problems of standards of materials and craftsmanship and, to some extent, to the related problem of the supply of materials. This could become obvious if the conservation officer was asked to comment when a roof failed unexpectedly early. Was the thatcher incompetent, or the supplier of materials, or was the quality of materials so impossible to judge in advance that neither could be held responsible? On the whole, it was the owner that was left to pick up what could be a large bill, although some thatchers have re-thatched roofs, at their own expense, in order to maintain their reputation, without knowing why they have been failed by material with every appearance of being sound.

In addition to the visual change that a re-thatch can bring, it is widely recognised that thatched roofs are of major archaeological interest where they retain historic and even medieval thatch. John Letts has established the importance of the thatched roof as a major resource to archaeobotany and to agricultural history (Letts 1999). Evidence of former craft techniques can be found in old roofs, including methods of fitting that are no longer current, or are rare. As an understanding of the archaeology of thatch has developed since 1992, it has become plain that modern methods of re-thatching that strip early layers of thatch, either extensively or completely, are responsible for serious losses.

The speed with which thatch has been drawn into the orbit of conservation legislation has been very variable. In some areas there has been a time-lag of over ten years between numerous buildings being listed grade II for the first time, many of them thatched, and the gradual and on-going application of existing legislation to thatching. This has created real problems. In the context of thatching it has meant, in effect, that thatchers working on some listed buildings experienced a period of relative freedom from controls and were cocooned from the concerns of local planning authorities and the conservation officer. They had grown accustomed to being able to do things their own way, in what has been something of a commercial free-for-all, and where the most effective sales pitch has played a large part in determining what goes on a roof. It should not be forgotten that the experience of large parts of the industry as it was developing was of the Bureau arguing for changing materials and for 'modernizing'. The conservation impetus to look back and consider how the past might be better treated in relation to the present and future, contradicted the Bureau's outlook before the 1980s.

In the limbo between the rural resurvey and the gradual application of conservation policies to thatch, not only have there been a good many conservation losses, particularly the unnecessary disappearance of medieval roof structures and medieval thatch, but some thatching businesses have developed, with attendant investment and plans for the future, without considering that the situation might change. It is hard for individual firms to accept adjustments to their operations in accordance with local planning authority control, especially when they see that control is patchy (occurring in one district, perhaps, but not in its neighbour), and, by inference, random. A firm specialising in water reed thatching, perhaps with a considerable interest in importing it (not only for their own use, but as an agent to other thatchers), may find it difficult to price up for long straw work (in which they may not be skilled) when long straw is required by the local planning authority. They may find themselves saddled with material or orders for material that they suspect they may be unable to use or excluded from tendering for work.

Thatching firms suffered from the recession in the early 1990s, like any other firms involved in (if not defined as part of) the building industry. Some went out of business, most saw their profits reduce and for some the application of conservation policies that limited their ability to price up for jobs came at a particularly bad time. Few thatchers are placed to avoid work on listed buildings, unless they operate exclusively on new build or the 1930s Vernacular Revival houses in the suburbs of large southern towns. It should come as no surprise, then, that there have been protests when local planning authorities have implemented thatching policies, and asked for listed building consent in connection with re-thatching and/or re-ridging. Without a single trade body and with many thatchers falling outside any trade body, it has been hard to discover how representative of the whole industry the protests have been. Thatchers who have welcomed conservation policies as providing a level playing field for competitive pricing, or who simply recognise that, where legal constraints and guidelines exist, any craftsman must expect to work within them, have been less inclined to seek publicity for their views and do not provide such good newspaper copy as those in opposition.

However, the process of developing a local planning authority thatching policy is usually a slow one. Many authorities have wisely taken the trouble to discuss pending policy with local thatchers and amend it following negotiation (eg in Somerset). It could also be said that those thatching firms which have complained, sometimes under the guise of arguments about the relative merits of materials, that the nature of their business and the free market is being upset by conservation policy, have usually had a good deal of time to read the writing on the wall and adjust to changed circumstances, as business of every sort has to do. There is some evidence to suggest that this is precisely what is happening at present, with a gradual acceptance that there is a *locus* for conservation in the matter of thatching.

The variety of ways in which some local planning authorities have approached thatching is bewildering. Some are inconsistent and even contradictory from place to place. Thatch is not a priority in districts or counties that have very little surviving or where, perhaps, most of their thatch is on 1920s or 1930s buildings. The expense of finding out what is or has been traditional in an area with relatively little thatch is hard to justify against all the other concerns of a conservation officer. Some conservation officers, like some thatchers, are more interested than others in the thatching tradition in their district or county. It has proved possible to interpret the existing guidance on listed building controls in many different ways. The 'like-for-like' philosophy of replacement has encouraged some planning authorities to require a repetition of the type of overcoat or ridge that appeared when a building was last re-thatched, even if there is good evidence that in the 1920s, say, it was thatched another way and that way was commonplace in the area. Other planning authorities have sought to support a local tradition by encouraging a change back to a type of thatching for which there may be archaeological or photographic evidence. Either approach can be justified, according to how the legislation is interpreted or in accordance with conservation philosophies, but to thatchers it can seem as though arbitrary regulation is being applied.

The government produces circulars and planning policy guidance notes (PPGs) to assist local planning authorities and others in development and land use issues. Circular 8/87, produced by the Department of the Environment, current until September 1994, referred specifically to thatch, Appendix IV, Technical Digest on Alterations to Listed Buildings. The relevant text 'Alterations in Detail' was as follows:

> I. External Elevations
> 1. Any alterations or repairs to external elevations should respect the existing materials and match them in texture, quality and colour ...
> IV. Roofs
> 1. The roofline is nearly always a dominant feature of a building and retention of the original shape, pitch, cladding and ornament is most important.
> 2. Thatched roofs should be preserved as far as possible as they are important survivals of a craft and a very early vernacular type. The availability of grants for repair should be made known and a list of local thatchers compiled. Relaxation of the safety provisions of the Building Regulations may sometimes be possible for thatched buildings. It is most desirable that formerly thatched roofs, where the thatch survives under a later cladding, should be restored.

The advice in PPG 15, issued by the Secretary of State for the Environment and the Secretary of State for National Heritage, updated the advice in Circular 8/87 and is current at the time of writing. It was adopted in September 1994 and is more detailed.

> Thatched roofs should be preserved, and consent should not be given for their replacement by other roofs. Where medieval thatch survives with characteristic smoke-blackening on the underside, it should be retained *in situ* and overlaid. When roofs are re-thatched, this should normally be done in a form of thatch traditional to the region, and local ways of detailing eaves, ridges and verges should be followed. Re-thatching roofs that have lost their thatch will require a waiver of building regulations in most cases, since they may not be allowed within 12 metres of a site boundary, but local authorities should be prepared to relax this rule if it does not constitute an unacceptable fire risk to other properties. (Paragraph C29)

Although it does not carry the weight of guidance on planning policy (although the source, Christopher Brereton's *The Repair of Historic Buildings* [1991], is a statement of English Heritage policies towards its repair grants), more detail was given by English Heritage.

> As examples of regional appearance, East Anglian roofs tend to be in regular planes, with steep pitches, pointed gables with rolled verges and often with decorative ridges, etc, while in the West Country shapes are softer in outline and lower in pitch, with half-hipped gables with rounded verges, and a general lack of ornamental features. In order to ensure that such regional features are not lost when re-thatching, an experienced thatcher should be employed who works in accordance with local tradition. Old photographs, etc may be consulted in order to ensure accurate reinstatement of original features. (Brereton 1991, 26)

There are worries that both types of straw thatching may disappear unless special efforts are made to foster their use. Some local planning authorities have made efforts to use constraints and encouragement, in the form of requiring listed building consent and providing information and grant aid, to preserve types of thatch over others and more are following suit. The situation is still very patchy. Meanwhile the industry has been surprisingly slow to respond to the opportunities that conservation represents. Firms and individuals comfortable with conservation have not advertised themselves as 'conservation thatchers' or set up as 'traditional thatch patchers'.

The problems at present, for thatchers, owners and conservation officers, are the need to understand the vexed question of what is 'traditional' in different areas and the need to improve the quality and availability of thatching materials. Well-researched publications of regional traditions are urgently needed, focused on answering the questions of what conservation controls are sensible, how they should be applied and where flexibility is essential.

Efforts to improve the quality and availability of thatching straw and extend the existing stock of managed reed beds could shift what was a quarrelsome debate in the early 1990s about the rights and wrongs of any conservation constraints into a much more fruitful area, that would help to ensure a healthy balance between the three common thatches and ease, for thatchers, the conservation policies that respect the history of local distinctiveness.

The importance of managed reed beds to wildlife conservation is an added impetus to increase them in extent and number. In spite of the cultural and economic value of the thatching industry today, the demand for thatching straw relative to that for grain is so tiny that there seems little chance of tempting breeders to create new varieties that would fit the needs of the thatcher. Letts, however, has shown that restoring types of wheat that were used in the past could be a way forward, using seed from gene banks. Current trials at the University of Reading are a huge step forward in the practical identification of suitable early varieties. If there are problems of husbandry and harvesting associated with cereals that have not been grown commercially since the nineteenth or eighteenth centuries, it is only practical experience that will identify what they are and how they might be overcome.

With a better understanding of regional traditions and better-quality straw and water reed available there would be more room for attempting to extend the range of the thatches again, in line with information on the diversity of thatching before 1940 that is becoming available, particularly in the work of Moir and Letts.

Part V

Appendices

Appendix 1 Thatching survey, 1965

A report on a survey conducted for the Rural Industries Bureau to assess the value of thatching as a national asset, on the present state of the craft and factors affecting it (1965) (private archive). Text in square brackets annotates the original.

INTRODUCTION

1. Thatch as a roofing medium has been a feature of the English countryside for four hundred years or more, and is probably the oldest of all building crafts practised in the British Isles.
2. Early thatching was probably a form of straw bonded by mud, but later developments in technique and material have resulted in the highly skilled craft and attractive architectural feature it is today, the standard being higher than it has ever been in history.
3. This is true of the craft as a whole, albeit that practices vary widely between various parts of the country. For example, in Hampshire and Wiltshire straw thatching is, as a rule, as thin as 5" [130 mm] to 8" [200 mm], whereas in the eastern counties it is applied as thick as 15" [380 mm], and for a skilled observer it is possible to identify a county or even an area by the techniques and the medium used.
4. Thatched cottages, inns and even manor houses form part of most people's mental image of the typical English landscape. They are pretty, have an air of history, fit neatly into the rural scene, and look delightfully old-fashioned.
5. But is a thatched roof old-fashioned? The craft is certainly ancient, but the artefact is as modern as the brick, the clay tile and the timber frame, and as a roofing medium has much to commend it.
6. Support for, or confirmation of many aspects of this report are contained in contributions from Mr Leslie Ginsburg, AADipl, S of P Dipl, ARIBA, MTPI, Head of the Leverhulme Department of Planning & Urban Design of the Architectural Association School of Architecture, and Mr J Ray, Research Officer, British Travel Association. These appear as Appendices A and B respectively [not reproduced here].

ECONOMIC VALUE

7. The intrinsic value of thatching to the English countryside cannot be measured in economic terms. The British Travel Association has been accused of using pageantry, 'tea cosy' cottages and the national character as gimmicks in its advertising campaigns to attract the tourist to this country.
8. Lord Geddes has ably defended this attitude in asserting that if this is what we have to offer then any good salesman will use these features in his sales campaign. (see Appendix C) [not reproduced here].
9. This technique can only be described as having been successful in the light of the £308,000,000 earned for Britain from tourism in 1964, which sum represented 7% of the total exports of Great Britain, the fourth largest single export, and the first dollar earner. It is not possible to allocate any portion from this sum as having been earned by thatch, but as this is a singular feature of the English countryside without which it would be very much the poorer, there is no doubt that it has made some contribution to this handsome return.
10. Apart from this aspect of the economic value, there are three others which perhaps can be measured in more real terms.
11. The first and most important is that over 600 thatchers are kept employed in the English countryside where they live, work, play and form part of the community, and whose children are educated and nurtured to the country life, and in toto it is estimated that these thatchers each earn on average at least £1,250 per annum, and spend between them, with the butcher, baker and candlestick maker in the countryside, the sum of £750,000 per annum. Add to this the total purchases made by them in pursuit of their trade, probably between £1 and £1.25 million, and a total sum of close to £2 million is indicated as being spent annually by them, almost entirely in the countryside.
12. The second is in respect of the money spent in the locale by visitors who, but for the attractions, would probably have gone elsewhere.
13. The third is in relation to property values; private property with thatched roofs command high prices, or need large sums spent on them; thus they are usually bought by more prosperous people who, having restored them to their original character, reside in them. This type of owner-occupier brings new life to the countryside and stimulates its economy.

HISTORIC VALUE

14. The yeoman cottages of feudal English are part of the heritage of this country and where good examples remain they create an atmosphere so often true to the age for which they were built, providing a glimpse into that period of history to which they belong.

15. Study Group No 8 of the Countryside in 1970 Conference referred to our 'Countryside Treasures' and wrote of them as natural, man-made or allusive, and while thatched housing is not specifically mentioned, neither are the manor houses, coaching inns nor a great number of other significant features of the countryside; one feels however that these were the things the Study Group had in mind when making its report. (P.8.3 para 1, P.8.11 para 45 (iv)(a))
16. The historic value of architecture featuring thatch is as real to the countryside as the Beefeater's uniform is to the Tower of London. Both could be dispensed with, but only detrimentally.
17. There is both allure and allusion in the cottage of Anne Hathaway at Shottery, Warwickshire, and the Barley Mow at Clifton Hampden, Oxfordshire, the scene of more than one of Jerome K Jerome's stories, and there is no doubt that places such as these will be maintained for posterity, primarily because they are associated with famous names, whilst others of equal architectural, historical or amenity value may be allowed to decay.

AESTHETIC AND ARCHITECTURAL VALUE

18. These are amenities that have been enjoyed for some four hundred years, but of which the modern masses only now, because of new-found mobility, are becoming aware. Many of our villages are excellent examples of community living, a mellow composition of colour and grouping that might have been designed by a talented planner.
19. But no expert has been at work; the harmonious blend of medieval church, Georgian manor house, half timbered inn with thatched roof cottages, is the product of long centuries of unplanned growth which owes its form to the economic requirements of the countryside of the past.
20. The British Travel Association uses these features extensively in its beautifully produced coloured literature (examples attached) [not attached], designed to 'sell' the English countryside, both at home and abroad, and it is not difficult to visualise the entirely different and less attractive picture which would be present without the Little Combertons, Selworthys, Broadhemburys and North Boveys.

THE PROPERTY SITUATION: AN ASSESSMENT [R E CLARKE AND G DALLISTON, THATCHING OFFICERS, RURAL INDUSTRIES BUREAU]

21. **Renovation**
    Much of the renovation and modernisation of thatched properties has resulted from changing circumstances over the years. because of this it is necessary to categorise the types of property:

(a) *Estate and Farm Workers' Cottages*
In the past these cottages went with the worker's job and an only be described as a roof over their heads, being very small with no amenities, little comfort, and of poor quality. This was acceptable to labour when it was cheap and plentiful, but with the drift from the land and the advent of the council house the agricultural worker now demands a better type of cottage. The result is that single cottages have had large extensions added, or two small ones knocked into one, or three into two, and then brought up to modern standards. To give these cottages the required amount of light, new windows had to be let in, involving extensive roof work, so that by the time they were finished and re-thatched they became nearly new buildings. Thus fewer cottages remain, but these are of much better quality, and had this not been done their probable end would have been demolition.

(b) *Redundant Estate and Farm Cottages*
Changes which have taken place on estates and farms have made redundant many of the workers' cottages, which have been sold to private owners or developers. These have been treated much as those above with more lavish fittings and decor, and in many cases with a completely new roof of Norfolk or wheat reed. Again there is a reduction in numbers but improvement in quality.

(c) *Properties Occupied by Retired Employees*
It was common practice years ago for employees of large estates to remain in their cottages on retirement, paying only nominal rent. Rising costs made these cottages a liability and maintenance was neglected. To do the minimum of keeping them dry, the roofs in many cases were stripped and galvanised, and sometimes a galvanised roof was laid over the thatch. As the old people died, these properties were sold as so much needed to be done to them, and in most cases they have been restored to thatch after modernisation by the new owner-occupiers.

22. **Properties which should have been Thatched**
Many thatched roofs have been lost through failure to put the work in hand soon enough. Too often it has been left until no longer weatherproof, or a property with a bad roof changes hands, when the thatcher is expected to come at very short notice. As most thatchers are heavily booked this is not possible, so in desperation the roof is stripped and an alternative covering applied.

23. **New Houses**
During the last few years thatchers have been increasingly thatching completely new large houses which were planned and designed for thatch, and thatching large extensions to existing properties. It is well known amongst thatchers that in the right environment an increasing number of houses are being designed for thatch, and this is likely to continue.

24. **Retention of Property Value**
Estate Agents will confirm that thatched property in good condition always finds a ready market; even old derelict cottages soon find a purchaser if they come on the market reasonably priced, and suitable for conversion and modernisation. Because a thatched cottage is 'something different', having character of

its own, and often historic and local interest, it commands a better price than other types of property. In most cases, owners of thatched property have a very real love for it in spite of the fact that with the increased cost of thatch and fire insurance it frequently represents a serious financial burden.

25. **Builders and Thatch**
Builders are not very helpful when dealing with thatch, often persuading the owner to remove the thatch rather than renew, in order to get the alternative roofing job. This occurs in many cases because a large number of thatchers will not undertake work through builders as they tend to add considerable sums to their estimates for little work or risk, so giving the impression that thatching is more expensive than it really is. Many purchasers know nothing about thatch or thatchers when they move into a district and often receive wrong advice with the result that the roof gets stripped.

26. **Thatch Decline**
Several years ago there was a rapid decline in the number of thatched roofs both residential and agricultural. In the case of residential property the decline has been halted and the modernisation of old properties and the building of new is enabling thatch to hold its own, although due to the change in farming trends many farm buildings are being demolished as they are no longer required, are unsuited, or wrongly sited for modern farming requirements.

THE MATERIAL SITUATION: AN ASSESSMENT. F W COOPER, THATCHING OFFICER, RURAL INDUSTRIES BUREAU

27. **Long Straw**
A few years ago it was an easy task to indicate certain clearly defined areas where one particular thatching material was used predominantly or even exclusively. For example, it was possible to follow a belt of long straw thatching running westerly from Chelmsford in Essex, through Hertfordshire and Bedfordshire, and thence through Northamptonshire and parts of Oxfordshire, north-west to Worcestershire and Herefordshire.

28. Throughout such an area, to which might be added the Suffolk/Norfolk border and other similar pockets in Hampshire and Wiltshire, property owners and thatchers thought mainly in terms of long straw thatching and relied on the local farmer for a supply of good straw grown from a recommended variety of wheat suitably harvested with a self binder and threshed with a threshing machine.

29. **Combed Wheat Reed**
Similarly, a study of the practice in the counties of Somerset, Devonshire and Dorset would have revealed combed wheat reed as the predominant thatching medium - Devonshire being its county of origin. This material is virtually the same as long straw, except that it is passed through a machine known as a reed comber which is fitted to the top of an ordinary threshing drum, and which removes the grain and the leaves from the stalk without the straw passing through the drum. The straw thus comes from the machine undamaged and with the butts all laid in one direction, and at this stage becomes known as wheat reed. Any other material would have been regarded as an intrusion upon tradition.

30. **Norfolk Reed**
The situation described in 28 and 29 above no longer pertains, the trend now being towards the more extensive use of the much more durable Norfolk reed, regardless of location of tradition.

31. Only careful observers and those having access to appropriate records will be aware of this transition, and some reasons for it are given below.

32. **Cottage Property**
One obvious factor is the ever-increasing amount of cottage property which is changing hands and being restored, often with new extensions. An encouraging number of new houses, as well as some previously tiled ones, have received a thatched roof; and as records show, Bureau Thatching Officers have dealt with a mounting volume of enquiries from architects, builders and property owners, most of whom have been looking for a material more durable than long straw or combed wheat reed.

33. **Combine Harvester**
Another factor bearing strongly on the trend towards Norfolk reed is the almost universal use of the combine harvester.

34. Each harvest sees the disappearance of a few more self-binders, and new methods of intensive husbandry dictate that straw will not be sold. Consequently there is a corresponding decrease in the supply of long straw and wheat reed, and three categories can be defined to which these sources of supply belong.
(a) the farmer with a small acreage of wheat who does not use a combine,
(b) very light or fen land where the use of the heavier combine harvester is impracticable, and
(c) farms where selected varieties of wheat are harvested with the binder, principally with thatching in mind.

35. **Growth Regulation**
A 'growth regulator' known as 'Cycocel' has been used experimentally in Suffolk this year. Sprayed on the wheat crop in early summer it reduces the length of straw by six or seven inches [150–175mm], the advantages of which are immediately apparent when harvesting with a combine.

36. **Impact**
With the decline of long straw and combed wheat reed, Norfolk and marsh reed will be increasingly used, and the demands made upon the tidal reed beds will be very heavy.

37. The reed producing potential of the marshlands is more than enough to meet the needs of all thatchers in the country; but emphasis must be laid on the word 'potential', for except in isolated pockets reed cutting is completely unorganised.

38. Reed growers could make a valuable contribution to the industry if they would be organised into a national association, for which there is every justification since Norfolk reed must be regarded as a national product in that it will, within the next ten years, become the main medium for thatching.

39. **Marshmen**

    It would be difficult to place in order of priority the many needs of the reed producing industry, but that of finding experienced marshmen and an adequate labour force during the cutting season requires urgent consideration. Young men must be attracted to marsh work, and some form of training scheme promoted.

40. **Reed Cutting**

    With mechanization on the increase much of the hard work can be taken out of harvesting, but there is need for training in cleaning, dressing and stacking the reed in marshes.

41. **Stock Piling**

    There have not been many seasons in recent years when reed supplies have been seriously affected by adverse weather conditions, but with the position as it is today, crop failure could have a very adverse effect, especially as the trend is towards Norfolk reed exclusively. A National Association could regulate stocks.

42. **Exporting**

    Stocks have occasionally been seriously depleted by the export of reed, and unless the annual surplus allowed, an Association could regulate exports in the interest of the national requirements.

43. **Annual Requirements**

    There are over 600 practising thatchers in England and Wales, and although there is a steady intake into the trade, it is not enough to take care of wastage during the next few years due to retirement, and those who fall by the wayside in their reluctance to learn a new technique, but if the numbers are maintained at the present level, the estimated annual requirements of reed will be 2,300,000 bunches. If nothing can be done to stimulate supplies of other mediums and thatching becomes dependent entirely on marsh reed, 6,000,000 bunches will be required, and to achieve this, careful systematic maintenance and development of reed beds will be essential.

44. A thatcher lays approximately 10,000 bunches per year, and as one acre on average yields approximately 400 bunches of reed, the total yield from some 15,000 acres would be required annually, indicating on this item alone an expenditure of approximately £525,000 per annum.

45. There is no question of this acreage being denied to other crops as such land is unsuitable for other cropping, whereas farmers owning reed producing land could, with good management, receive an adequate return from reed harvesting. Various estimates indicate an average of £30 per acre net.

Appendix 2   A trainee thatcher reflects on her trade in 1950

The following (Public Record Office 16) is a rare piece of evidence for thatch training for the disabled and a reminder that women thatched during and after the Second World War. It is also an important, since rare, documentation of the role of women workers in agriculture and the building trade.

By 1943 Agricultural Committees were organising training for women thatchers (London Evening Standard 21st October 1943). Ruth Pollock's account is also instructive in identifying the difficulties of an 'outsider' learning from a masterman, and in listing the numerous tasks, apart from putting straw on the roof, that any thatcher has to undertake.

Letter from Ruth Pollock (a former trainee thatcher) to the Ministry of Labour and National Service, 26th October, 1950.

Dear Sirs,
As I have now finished my training as a building thatcher under the disabled persons training scheme, I should like to express my gratitude to you for the training I have received and my appreciation of the scheme in general. Thatchers, and particularly thatching trainees are, however, comparatively rare, and as their conditions of work are somewhat different from other crafts and trades, I wondered if you might be interested in some comments on the scheme from the trainees' point of view.

With regard to the time allowed for this craft, I feel that 13 months and 12 months practical work with the tutor should be quite sufficient, but in many cases this may not be so, through no fault of the trainees. Short and long tail wheat straw thatchers in particular frequently spend almost all their lives working on their own with perhaps one helper who can yelm but do little else. They are, therefore, unused to analysing their own work or comparing it with anyone else's so it is sometimes difficult for them to know what it is they want to teach and how exactly they do it when they have discovered. In Norfolk and Devon reed thatch, the thatchers work in teams and so this difficulty is much less marked. Further there is a tradition amongst certain thatchers which fortunately I believe to be dying out, that they must not give away their best secrets and, therefore, they do not tend to teach in the ordinary sense of the word and the pupil has to watch very carefully to find out for himself.

Another unusual feature of thatching is that in many cases the teacher and trainee work together for months on end and seldom see anyone else except for casual greetings. This means that unless they get on fairly well together the situation rapidly becomes untenable and so it is extremely desirable that the trainee should go and spend at least one day with the thatcher at work to see what the conditions are before they agree to the training. Even so in some cases they may not be able to agree for the whole period of training, although the trainee may be learning well and make an excellent thatcher later. But how these friction between personalities could be lessened it is difficult to see except by the persons affected.

The travelling allowances are a very helpful and welcome provision, indeed they would be essential to most thatcher trainees, for the majority of thatchers work over a very wide area and the work is not often near base. This applies to all types of thatching, if the trainee is to have experience of the various types of difficulties he will encounter later. It is impossible to find all the various shapes and conditions of roof in one place and each has its own problems. A liking for change in surroundings is an essential quality in a thatcher who must be prepared to put up with the accompanying difficulties such as lengthy journeys to work and sometimes periods in lodgings. It is not a job for someone who likes to plan far ahead, travel to work on a regular bus or live on the job. He could, of course, solve the latter difficulty by living in a caravan but this is not everyone's choice and it is not always possible to find a suitable site close at hand.

When I came to get my tool kit I was in a quandary as I was not able to follow the Ministry of Labour's list exactly and obtained what was essential. I suppose this is because each type of thatching requires slightly different tools. I did not in fact exceed the money value allowed although with the rising cost of tools this may not long be possible.

I use a saw almost every day, mainly for reducing wood to suitable lengths for splitting, and this tool was not mentioned. It is not possible to do this work with a hood as it sometimes splits the wood and so one loses several inches, and the piece becomes unusable. Further no eave-trimming knife or hook is allowed although this again is used by all the different types of thatchers. The list includes a 'needle with a curved handle'. I have seen two distinct types of needle in use but neither have a handle, which would prevent it being pulled through the thatching when necessary and so make it most awkward to use. Another essential tool not mentioned is a pair of wire cutters. A thatcher often has to remove and replace wire netting with which some roofs are covered. We also use a rough and a fine stone as the tools used for cutting straw have a rougher edge than those used on the wood. The leggettees I did not obtain as I believe they are used in reed thatching, which I have not yet learnt, but which I hope to learn as soon as I am really proficient in short tail. Then a chisel is necessary for removing old metal or cement from around chimneys so that they may be made watertight, and a bricklayers trowel for packing cement round them where they is no metal, and where pointing is necessary. I am aware that some thatchers use other means for sealing off the chimney stacks, but this is the one preferred by my teacher and as it is satisfactory and economical it is, therefore, the one that I shall normally employ.

When I started my training there were many people who said that I would 'never stick it', and 'it would be far too heavy' and that it would be 'dull' for anyone 'with a certain amount of intelligence and education'. It is a heavy job for a woman and, therefore, requires good physique but it is not heavier than certain types of domestic, laundry or farm work which are regularly done by women, and it has the advantage of being contract or piece work which enables the thatcher to work in his or her own time and so eliminate unnecessary strain. (It is also important because they can stop roof work when a wet and slippery ladder or a high wind makes it dangerous). A woman's lesser physical strength does mean, however, that although she can earn a reasonable wage she can never earn quite as much as an agile man on the same contracts unless she is a better thatcher. It also means that a man with certain types of disability could be a successful thatcher.

The qualities required in someone taking up thatching I should say were: integrity, perseverance, patience, physical strength, and if not a knowledge of drawing at least an ability to see a straight line and judge an angle. Particularly if he is going to work on his own he must be able to improvise and be willing to turn his hand to many jobs which at first sight appear to be outside his province but which would lose him contracts if he did not tackle them. Tact is another obvious asset when he is working for people whom he has not time to get to know and when he has so many opportunities for getting in an occupier's way. If he is working on an estate or for a builder there is, of course usually someone to help him out, and the timbering of a new roof, tying and loading of straw, and cutting of wood will often be done for him. These are examples of jobs which he must learn to do outside thatching (timbering for tiles is not the same) and which it is not easy to learn to do well in the time available as there are not enough opportunities for practice.

To return to my list of qualities essential to a good thatcher I have deliberately listed them in this order as without the first three integrity, perseverance, and patience, the work must suffer, will not last and so jeopardise his chances of future employment.

From this it will be seen that it is not always the most obvious people who will make the most successful thatchers, and that it might be particularly suited to certain types of disabled or sick people (particularly where rhythmic physical work will help) and perhaps displaced persons, as the language difficulty is not so important. I think it should appeal to the individualist looking for independence together with some creative satisfaction and especially if he likes solitude and quiet when working. Certainly 'a certain amount of intelligence and education' are not a handicap but a great advantage.

Appendix 3 Historiography

Publications on thatching immediately before the Second World War and from 1945–94 reflect changing perceptions of thatch. The following sample illustrates something of the range and content of what has been available in print. Thatching straddles both agriculture and building. This has emerged in the uncertainty as to where to categorise thatching at a bureaucratic level and is reflected in the kind of publications that have included accounts of its history.

As with thatching itself, the Second World War seems to have been a break in the continuity of writing about the history of thatch. Between the wars, and particularly in the 1920s and 1930s, books on the English countryside and country crafts were popular, and a number included chapters or sections on thatching. In many cases these were based on conversations with practising thatchers. The outstanding accounts of this kind are to be found in Hennell's wonderful *Change in the Farm* (1934), and Hartley's *Made in England* (1939), both with fine line drawings illustrating aspects of thatching (Figs 33 and 34). Both texts lead from rick-thatching to house thatching in a way which was natural at the time they were written, when straw thatchers did both. Both texts include information from discussion with and observation of actual thatchers at work. Hennell's text is important for information about harvesting and threshing and his sections on house-thatching are particularly interesting. He attributes the introduction of fancy detail on houses to the Farman firm.

The fashion of herring-boning in place of diamonding, of making other fancy patterns of split hazel runners and of working the edges of the outer layers of reed into ornamental shapes has been introduced in the last fifty years by the family of Mr Farman of Salhouse near Wroxham, who is one of the best-known practitioners of the craft. (Hennell 1934, 24)

Unfortunately it is difficult to be sure from this account whether the author is including the scallops and half diamonds of the ornamental block-cut ridge here, or whether he is referring to the orders of cut pattern between eaves and ridge which are sometimes seen both on long straw and water reed roofs.

Hennell acknowledges that his description of technique came from a Dorset thatcher. He refers to local uses of different materials for thatching houses: rye straw in Somerset, for instance 'grown solely for this purpose' and flax used (on a couple of cottages) in Rougham, Sussex. Hartley, too, refers to different local materials, including oat straw, rush and ling heather. She also refers to some interesting techniques, notably the practice of throwing old fishing nets over the thatch and pegging them down as a protection against gales and birds, perhaps the origin of wire netting. There is archaeological evidence for the use of fishing nets over thatch in Cornwall. Hartly also mentions large gauges of wire netting used under the thatch for 'patching and contriving' (Hartley 1939, 53)

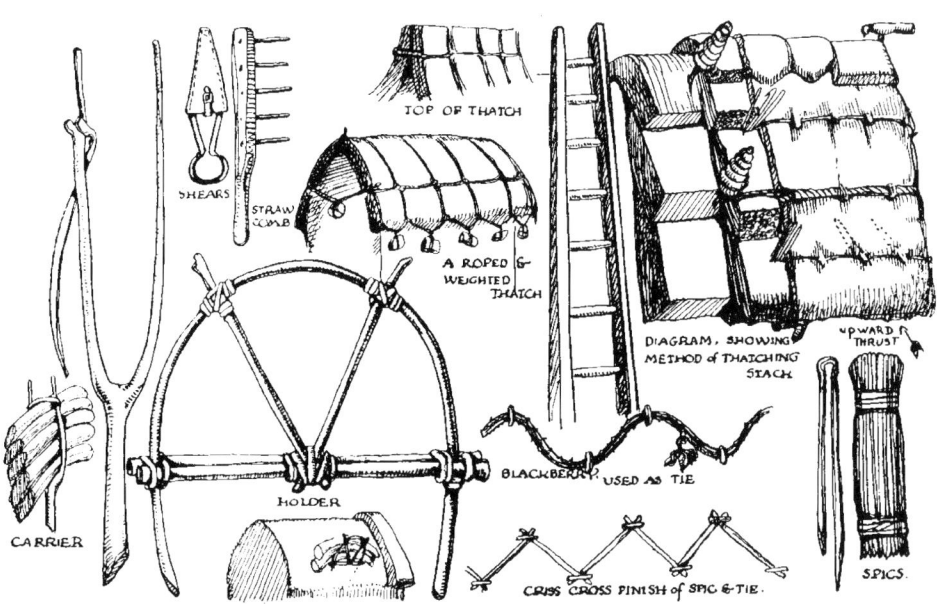

Figure 33 Thatching tools and methods, from Hartley 1939, 61.

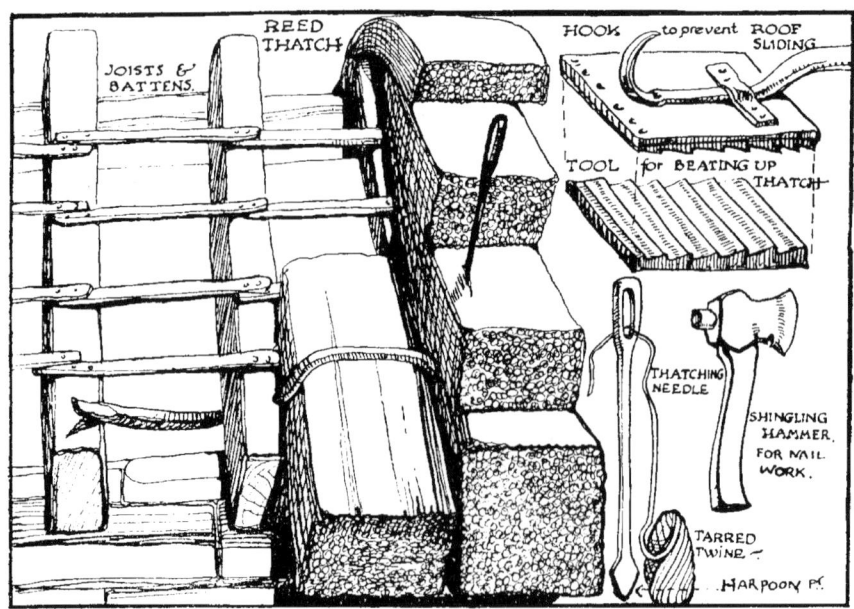

Figure 34 Thatching impedimenta, from Hartley 1939, 56.

and a method of plastering the ridge which was still current in some areas. Methods of ridge-plastering with mud were recorded by J R Harrison during field studies carried out in Cumberland between 1979 and 1982. Harrison also noted the survival of the technique as ridge copings of cement and sand in Northamptonshire in the same period (Harrison, 1989).

Hartley's section on house thatching is illustrated with photographs showing a water reed thatcher at Lydd in Kent, who ridged in straw, with wire netting over the ridge only, and who intended to cover the remainder of the roof with tarred fishnet, once it had 'settled'. She also illustrated a long straw thatcher at Malmesbury in Wiltshire, using blackberry liggers.

Few other texts of the 1930s on country crafts or agriculture contain more detail on thatching than Hennell and Hartley. Batsford and Fry's *The English Cottage* (1938) sums up the state of knowledge and makes a plea for a more comprehensive account of local tradition, which sadly, no writer seems to have taken up after the war, when the evidence was more marked than today.

> If much has been written in scattered and sporadic fashion about thatchers' methods, it is essential that some thatch-lover should do what has not yet been attempted, ie work out a careful comparative account of the various local types of thatching - how, for instance, the East Anglian style differs from the thatch of the South-west, how the gable ends and roof ridges are treated, how the roofing is brought over and round the dormers, how hips and valleys are handled, etc. Little can be done here to define these local distinctions; for example the eastern roofs are of steeper pitch, and the thatch is cocked up with perky sharpness over the dormers. The South Western thatch has a peculiar velvet texture, loves wavy lines and wanders round corners. About Ampthill (South Bedfordshire) the dormer thatch is cut with particular sharpness almost into a complete semicircle. Wiltshire cottages have their own dormer thatch-shapes, and often each porch of a row will have its little hat of thatch. The mud or cob walls have also their own little roofs of thatch in Wiltshire and the adjacent counties. (Batsford & Fry 1950, 75–6)

Books on buildings, rather than agriculture or crafts, are scanty in references to thatch in the 1920s and 1930s. Nothing can compare with Innocent's earlier account in *The Development of English Building Construction* (1916, 188–222), which must stand as the one of the best published historical accounts to date. Even at that early date he was talking of the loss of regional variation in the face of improved communications and what he described as 'the capitalist thatcher', and complaining that the 1910 thatching pamphlet produced by the Ministry of Agriculture and Fisheries was likely to 'destroy those local aspects of the craft which make it so interesting to students of old building construction' (Innocent 1916, 188).

After the Second World War no-one took up Batsford and Fry's suggestion of a book dealing with local variations, and references to these, which continued to be made in the Batsford publications, did not develop the theme. One Batsford book, Wymer's *English Country Crafts,* published just after the war (and perhaps researched before the war) considered that the local production of materials was the basis of regional variation.

> Each district uses only locally grown materials – reeds, wheat, rye or oat straw, or heather, as the case may be – so that a roof thatched in one area would still invariably have an appearance quite unlike that in another village even if both were done by the same man. In Devon there is the unthreshed wheat straw, grown in the north of the county and known as wheat-reed; Somerset thatchers rely on rye straw; in Essex rushes are used; while in East Anglia we have the famous Norfolk reed, possibly the best of all; and so on. (Wymer 1946, 50)

These references to a variety of materials located in particular areas were something of a rarity after the Second World War.

Salzman's *Building in England Down to 1540*, completed in 1934, first published in 1952 with a revised edition of 1967, must be mentioned. As a scholarly account of pre-Reformation references to the building and allied trades, the section on thatching is unlike any other previous text and was the first detailed attempt to understand medieval thatching from documentary sources which were often highly technical, legible only with difficulty and produced when variant spellings of terms that may have differed from place to place, were legion. Salzman deals with thatch in a chapter that includes other roof coverings which is useful for context and a grasp of competing materials. This pioneering publication was the first real glimpse into the diversity and extent of pre-1540 thatching. It includes documentary reference to different thatching materials, different kinds of thatchers and different items that might temporarily be protected with thatch, for example the engines of war at the Tower of London, as well as roof thatches.

The major post-war text on thatching was a complete departure from previous books and was *The Thatcher's Craft* (Morgan & Cooper 1960) produced by the Rural Industries Bureau (revised in 1977 for the Council for Small Industries in Rural Areas). Text is not really the correct word, since *The Thatcher's Craft* is really a book of photographs with long captions. It has become the classic reference book on thatching, produced by practitioners and stripped of 'romance'. The book shows the three common post-war thatches being fitted to model roofs by an operator in the manner of a freeze-frame exhibition or sequence of processes in a factory, resulting in a presentation of thatching as colourless 'technique'. The book was designed in the first instance to help trainee thatchers, as well as 'architects, builders and householders', as the foreword by John Betjeman states.

The objective seems to have been to show thatching in the manner of a modern building trade, the method in the book having parallels in 1950s textbooks on brickwork or carpentry which showed different bonds being laid, or methods of constructing windows. The book drew thatching out of the folksy realms of nostalgia for a country past and into the mainstream of building trades for which there were apprenticeship schemes and training courses. The Bureau, of course, had an interest in defining it in this way and establishing its accessibility to contemporary architects. The continued popularity of the book reflects its attractive step-by-step explanations of the processes of thatching which, before, had been surrounded with mythology and an absence of plain information. It did, however, underplay the complex business of the production and supply of materials, which were pervasive themes in the Bureau's own annual thatching reports. This aspect of thatching did not merit any photographs in the book and was pushed into the background by the emphasis on fitting technique.

There is very little in *The Thatcher's Craft* on the relationship between the thatcher and the production of his materials, which the popular 'agricultural history' books were bound to include. Moreover, the Bureau was virtually silent on the matter of regional variation. The first chapter gives two pages of history, mostly pre-1540 and drawn, with acknowledgement, from Salzman, and notes that the development of water reed beds (which the Bureau was promoting at the time) outside Norfolk would help to offset 'the loss of the familiar wheat straw which has long been characteristic of thatching in Hampshire, Dorset and Devon' (Morgan & Cooper 1960, 2). The introductory sections to the three common post-1940 thatches make no reference to their common locations.

Alec Clifton-Taylor's popular *The Pattern of English Building* (1962) is more opinionated with regard to regional variation, but clearly influenced by *The Thatcher's Craft* in figures quoted for durability and in its attitudes to water reed. The author refers to a range of materials in use historically, then commends the Bureau for transferring combed wheat reed roofs 'which almost equal the best Norfolk thatch in appearance', from the south west further east. The author is more concerned with aesthetics than historic regional variation, which he attributes, not to the traditions of the locality, but either to the individual thatcher or to dynastic family style. This is a theme that was taken up by some thatchers in the early 1990s. They argued from it that conservation policy prevented the thatcher from expressing his or her creative individuality on a roof. Clifton-Taylor's preference is clearly for ornamentation, within limits. The individual thatcher 'really comes into his own' with the ornamental block-cut ridge, but 'used in a fussy way, curving round a proliferation of little dormers and gables, thatching can be rather tiresome'. Clifton-Taylor concludes by dismissing any concern that an increasing use of water reed might be inappropriate in certain localities.

> Does this offend against the *genius loci*? Is reed an unwelcome intruder into traditionally straw districts? Here the answer is an emphatic 'No'. Thatch with its quiet colouring (as soon as it has weathered) and, in the case of reed, its gentle texture, is visually so accommodating that, as suggested earlier, there is scarcely any material with which it does not look well in country places. Please, in country places: not on the by-pass road; not in the residential street (where it always looks arty-crafty); above all, even if the bye-laws permit it, not at the shopping centre (Clifton-Taylor 1962, 280–90).

In 1968 the Rural Industries Bureau did consider the idea of a professionally written history of thatching but this idea was rejected by the Development Commission on grounds of cost (Public Record Office 15).

Shire Album 16, *Thatch and Thatching* by Fearn (1972), in line with the theme of the series, includes useful information about methods of production, adding to this a selection of the same photographs produced in *The Thatcher's Craft* and clearly influenced by coeval thinking at the Council for Small Industries in Rural Areas at the time of publication.

Billett's *Thatching and Thatched Buildings* (1979, revised and enlarged edition 1988) includes general chapters followed by a description of a selection of thatched buildings in different regions. The publication reflects the difficulties of writing with authority on thatching, given that performance issues are difficult to understand and regional variation hard to explain, without a good understanding of the history of the supply of materials. It is only too easy to mix useful information with misleading statements or inference, for example that the pitch of a roof most likely to deteriorate first is the one that receives less sunshine, or that the tighter thatch is packed, the more durable it is. While there are always exceptions to the rule with thatch, it is generally the pitch of a roof most subject to extreme changes of temperature that deteriorates first. Research at the University of Bath established that there is an optimum for the tight-packing of thatch, and that tighter packing can accelerate decay by reducing ventilation. The implication that listed building consent only applies to external changes to a building (Billett 1988, 87) could put owners of listed buildings on the wrong side of the law, and it is a pity that the advice of the Society for the Protection of Ancient Buildings was not sought regarding the wisdom of introducing damp-proof courses in mass wall buildings. The lists of thatched buildings in different regions, while a good guide to where concentrations of thatch can be found, does not always mention what sort of thatch they are covered with.

Billett followed the first edition of *Thatching and Thatched Buildings* in 1984 with the first whole book devoted to thatched buildings (rather than thatching) in one county, *Thatched Buildings of Dorset*.

Dorset is, historically, an especially complex thatched county, including an historic junction between the straw thatches, as well as native water reed but this does not emerge in the text. Billett's book includes useful information about thatching practice in the county in the 1980s but the historical background is not covered in depth. The thrust of the publication is a gazetteer of thatched buildings where the type of thatch is not always mentioned.

In 1986 the Society for the Protection of Ancient Buildings produced a technical pamphlet, *The Care and Repair of Thatched Roofs*, written jointly by Peter Brockett, then a thatching officer with the Rural Development Commission, and Adela Wright. This was not intended to be a history of thatching or an analysis of regional types, but in an extremely short span (12 pages) recognised the existence and value of both and provided owners, architects and conservation officers with a wealth of valuable advice. The impact of re-thatching on the different elements of an old building was covered in detail and a good portion of the pamphlet is designed to educate architects in how they might avoid pitfalls in new design for thatch. This was a remarkable publication when first produced, still is and should be required reading for the owner of any thatched property.

In 1987 *Thatch: a Manual for Owners, Surveyors, Architects and Builders* by Robert West, the proprietor of Thatching Advisory Services Limited, was published. The first chapter included some of the familiar medieval documentary references. This book recognised historic regional variation.

> An experienced thatcher if transported blindfold to any rural part of England would be able to tell where he was by the style of the thatch. (West 1987, 57)

Like *The Thatcher's Craft*, it placed a good deal of emphasis on defining thatch as a modern building material. It paralleled *The Thatcher's Craft* in its use of photographs of technique. The text extended the process of demystifying thatch by including advice on how to identify the three common thatches and very helpfully suggested how owners could carry out modest DIY repairs. In the chapter on 'Learning the Art of Thatching' West also sought to demystify training, arguing that it could be carried out successfully in a far shorter time than the old apprenticeship system.

> It has been accepted that the old form of apprenticeship is no longer reasonable or practical in modern competitive society. The alternative is to select an adult who has already had work experience, and who understands both the physical and mental pressures of a demanding job, and put him through a structured, intensive training course for six months. (West 1987, 128)

Thatchers and Thatching (1991) by Nash is largely occupied with Dorset, although the existence of Billet's book on Dorset's thatched buildings may have persuaded the author and publisher that the brief should be wider. This is a valuable contribution, written with the kind of familiarity with thatching that comes, not only from research, but from family connection with the trade. It includes useful documentary references to particular thatching projects.

Regional studies of thatching with an historical perspective are urgently needed and researchers should take up Batsford's plea of 1938. Moir and Letts have established that primary documentary sources do exist, sometimes in abundance. The survival of this material is patchy, but what does exist deserves examination and over time could be supplemented from references to thatching which crop up in sources, particularly leases, which will never be indexed in any archive under 'thatching'. Recent documentary research into one Dartmoor farmstead brought to light a lease which required the tenant to grow and keep in store a certain amount of combed wheat reed per annum for patching thatched buildings on the farm. The farmhouse has since been slated and the farm buildings have tin roofs, but retain the external timber pegs for hanging ladders which the farmer reports were used both for picking cider apples and for patching jobs on the thatch. For the period from 1940, for which there are fewer deposited records, oral history is an invaluable tool for understanding regional variation. The impact of large estates on the use of materials needs more investigation, as does the diversity of long straw techniques, both from locality to locality and within one locality.

Glossary

ANAEROBIC Without oxygen.
AWN A stiff bristle growing from the ear of the plant.
BARGE The inclined edge of a roof rising along a gable, as opposed to the eaves.
BROACH [brotch, brawtch etc] An alternative term for *spar*.
CARR Fenland with water of a neutral pH.
COMBED WHEAT REED [combed wheat, wheat reed]
1. A form of thatching straw composed of stems that have been combed mechanically to remove grain and extraneous waste material without crushing the stem. The modern form of *reed straw*.
2. The modern technique of thatching a roof with this material, using the material in bundles which retain a common orientation to the stems. The material is dressed into place and usually secured without external fixings other than at the ridge.
CROOK An iron nail with a hooked head used to secure thatch directly to rafters by clasping the *sway* that holds a course of thatch in place. Formerly used only in difficult positions on a roof, but now used routinely in straw and *water reed* thatching.
CRUSHED STRAW Term used in these volumes to denote straw that has been bruised along its length and towards its lower end by threshing, and which subsumes the modern material known as *long straw*.
DRAWING The technique of pulling stems of a heap of material, thus aligning them for use as thatch and removing short stems, leaves and weeds. Now associated with *long straw*, which is drawn into *yealms*.
FLEEKING The practice, and product, of laying a thin coat of *water reed* (or other materials) beneath the lowest layer of the thatch to provide an even surface.
HAULM [halm, helm] See *stubble*, and also *helm*.
HELM [halm, haulm] A term (now obsolete) for *reed straw* as used in southern England beyond the West Country down to the mid nineteenth century, and composed of stems crushed only at their tops during threshing. The relation of *helm* to *haulm* is not yet entirely clear.
LEGGAT [legget, etc] A bat with one surface treated to catch the ends of the straw or reed (in *reed straw* and *water reed* thatching), with which the bundles may be beaten up under the *sways* to tighten the coat. Not used with *crushed straw* or other flexible materials.
LIGGER A length of roundwood (usually hazel or willow), often split, laid over the upper surface of a thatch to hold it in place, with the help of *spars*, and therefore similar to a *sway* except for its position. Rarely used in modern *water reed* or *combed wheat reed* thatching except on ridges, but used to hold in the eaves and *barges* of a *long straw* roof.
LODGING The flattening or buckling of crops in the field by the action of wind and rain (or other factors).
LONG STRAW [longstraw]
1. One of the two principal forms of thatching straw now used in England (the other being *combed wheat reed*)and the modern version of *crushed straw*. This material was identified by this name in the early nineteenth century, to distinguish it from the shorter straw removed in the course of mechanical threshing.
2. The modern technique of thatching with this material. The straw, which has usually been harvested with a reaper-binder and mechanically threshed, is *drawn* out of a wetted heap into *yealms* before being laid on the roof; the straw therefore usually has a proportion of butts pointing up the roof. The roof is crooked into place or is secured to the pre-existing base coat by *spars* and *sways*, rather than dressed into place as in *water reed* or *combed wheat reed* thatching.
MONOCOT Monocotyledonous; botanically, a flowering plant with one seed leaf that often reproduces by rhizomes as well as seed; in thatching, a member of the grass, rush or sedge families.
NITCH A bundle of combed wheat reed, historically 28lbs (11.2 kg), but now sold in bundles half that weight.
REED STRAW Generic term for all thatching straw which, in contrast to *crushed straw*, has had its grain removed without crushing the lower portion of the stem. *Combed wheat reed* is a modern version.
RHIZOME An underground stem of variable thickness that produces buds which can grow into new plants.
SPAR A section of roundwood (usually hazel) split, sharpened, and twisted into a U-shape. Thrust into the thatch the spar holds one layer to another, usually by holding down a *ligger* or a *sway*. See also *broach*.
SPAR COATING The fixing of a new layer of thatch onto an existing layer using *spars* and (usually) *sways*. Persistent spar coating produces stratified accumulations of thatch.
STINGING [stingeing] The practice of inserting new material into a roof requiring repair, by opening a hole and forcing the new material into the gap.
STOB A small plug of new material inserted into a roof in the repair process called stobbing or *stinging*.
STUBBLE The uncrushed residue of the straw left standing in the fields after harvesting, which when tall could be mown and used for thatching. In this form also called *haulm*.
SWAY A section of roundwood, formerly of hazel or willow but now often of iron, used horizontally in combination with *spars* or *crooks* to clamp a new course of thatch into position. Each sway is concealed in the finished roof by the next course of thatch or by the ridge. See also *ligger*.
TILLERING Cereals produce tillers, lateral shoots from the base of the stem, during the winter and spring which eventually elongate and produce ears.
WALE Reed-beds cut annually produce 'single-wale'; beds cut every two years produce 'double-wale'.
WATER REED
1. Wetland plant (today *Phragmites australis*) used for thatching.
2. The technique of thatching with this material, which is carried onto the roof in bundles and secured butts down with *sways* and (today) *crooks*. The reed is driven up under the fixings with a *leggat* to tighten the roof.
YEALM The bundle formed in *long straw* thatching from handfuls of straw which have been *drawn* from the heap of threshed material.
YEALMING The practice in *long straw* thatching of forming the bundles called *yealms*, by *drawing*

Bibliography

This Bibliography is reproduced, with references added specific to this volume, from volume 5 of the Research Transactions, *Thatching in England 1790–1940*, as a valuable resource to thatch and other historians.

PUBLISHED SOURCES

Airs M, 1981 Hovels or helms: some further evidence from the seventeenth century, in *Vernacular Architecture*, **14**, 50–51.

Albino H H, 1946 House thatching, in *Somerset Countryman*, **15**:4, 68–71.

Alcock N W and Laithwaite M, 1973 Medieval houses in Devon and their modernization, in *Medieval Archaeology*, 17, 100–25.

Allingham H and Dick S, 1991 *The Cottage Homes of England*, London, Bracken Books.

Angold R E, Sadd P A and Sanders M, 1998 *Fire and Thatch. Project report for Partners in Technology Project Number: CI 39/3/2866. Specifications for Materials and Treatment of Thatch* Volume 1, High Wycombe, RHM Technology.

Angold R E and Sanders M, 1998 *Longevity of Thatch in Relation to the Surface Properties of Straw. Final Report for the Partners in Technology Project DoE reference CI 39/3/286* Volume 2, High Wycombe, RHM Technology.

anon, nd Time to remember: 100 years of local history, *Northampton Evening Telegraph* [press cuttings file, Northamptonshire Library].

–, 1804 *Results of the Enquiry (on Labour) being a Recapitulation of the Average Wages of Several Counties 1790–1803*, Board of Agriculture.

–, 1807 5th ed *The Complete Farmer or, General Dictionary of Agriculture and Husbandry: comprehending the most-improved methods of cultivation; the different modes of raising timber, fruit and other tree; and the modern management of livestock: with descriptions of the most approved implements, machinery and farm-buildings*

–, 1909 Thatching, in *The British Architect*, **71**, 145–6.

–, 1916 Thatching, in *Journal of the Bath & West Somerset Counties Society*, 5th ser, **XI**, 274–7.

–, 1922a Straw reed for thatching, in *The Architect*, **107**, 276.

–, 1922b A decaying industry, [unknown newspaper cutting, Northamptonshire Local Studies Library].

–, 1922c, in *Birmingham Post*, 31 August.

–, 1923a, in *Western Morning News*, 13 July 1923.

–, 1923b New cottages at Tetbury, in *The Architectural Association Journal*, **39**, 61.

–, 1926 The revival of thatching, in *The Studio*, **91**, 93–7.

–, 1927a Thatching by reedwork, in *The National Builder*, 34–35.

–, 1927b Modern Practice. I – Fireproofing thatch, in *The Architect & Building News*, **117**, 1066.

–, 1931 The present position of the use of fertilisers, in *Journal of the Royal Agricultural Society* **92**, 162–176

–, 1932 The ancient industry of thatching, in *The Builder*, **143**, 130.

–, 1933 The thatched roof, in *The National Builder*, 236–238.

–, 1948 Secrets of thatching must be saved, in *Northamptonshire Independent*, 18 June 1948.

–, 1949 Portrait of a country craftsman: 3: R W Farman, reed thatcher, in *The Village*, **1**.

–, 1953 *A Report from April 1952 to March 1953 on the Development of Skill in Country Workshops*, London, Rural Industries Bureau.

–, 1959a Porthleven's last thatched house, in *The West Briton*, 23 April.

–, 1959b, in *The South Devon Journal*, 29 April.

–, 1962 Running a Cottage to Earth, in *Ideal Home*, June, 31.

–, 1967 Old cottage is protected, in *The Guardian*, 22 March.

–, 1971 *Ideal Home*, July, 44.

–, 1990 *Het Weke Dak*, Den Haag.

Armstrong A, 1988 *Farmworkers: A Social and Economic History 1770–1980*, London, Batsford.

Bacon R N, 1844 *The Report on the Agriculture of Norfolk*, London, Chapman & Hall.

Bagshawe T W, 1951 *Disappearing Rural Industries, Trades, Crafts and Occupations*, 5 vols, unpublished PhD thesis, Bagshawe Collection.

Bailey C, 1856 *Transcripts from the Municipal Archives of Winchester*, Winchester, Hugh Barclay.

Bailey J, 1810 *General View of the Agriculture of the County of Durham, with Observations on the Means of its Improvement*, London, Board of Agriculture.

– and Culley G, 1794 *General View of the Agriculture of the County of Cumberland with Observations on the Means of Improvement*, London, Board of Agriculture.

–, 1800 *General View of the County of Northumberland*, Newcastle, Board of Agriculture.

Baker A E, 1854 *Glossary of Northamptonshire words and phrases*, 2 vols, London, J R Smith.

Banks J, 1807 On the culture of spring wheat, in *The Repertory of Arts and Manufactures*, 2nd ser, **10**, 42–48.

Batchelor T, 1808 *General View of the Agriculture of the County of Bedford*, London, Board of Agriculture.

Bateman S, Turner R K and Bateman I J, 1990 *Socio-economic impact of changes in the quality of thatching reed on the future of the reed growing and thatching industries, and on the wider rural economy*, Report to the Rural Development Commission, School of Environmental Sciences, University of East Anglia.

Batsford H and Fry C, 1938 *The English Cottage*, London, Batsford.

Baxter J, 1834, *The Library of Agricultural and Horticultural Knowledge*, 3rd ed, Lewes, J Baxter.

Beaven E S, 1909 Pedigree seed corn, in *Journal of the Royal Agricultural Society*, **70**, 119–139.

Bell J and Watson M, 1986 *Irish Farming, Implements and Techniques 1750–1900*, Edinburgh, John Donald.

Bibby C J and Lunn J, 1982 Conservation of reed beds and their avifauna in England and Wales, in *Biological Conservation*, **23**, 167–186.

Biffen R H, 1915 Spring wheats, in *Journal of the Royal Agricultural Society*, **76**, 37–48.

Biffen R H and Engledow F L, 1926 *Wheat Breeding Investigations at the Plant Breeding Institute, Cambridge*, Ministry of Agriculture & Fisheries Research Monograph, **4**, London.

Billett M, 1979 (1988) *Thatching and Thatched Buildings*, London, Robert Hale.

Billett, M, 1984 *Thatched Buildings of Dorset*, London, Robert Hale.

Billingsley J, 1798 *General View of the Agriculture in the County of Somerset with Observations on the Means of its Improvement*, 2nd ed, Bath, Board of Agriculture.

Bingham J, 1979 Wheat breeding objectives and prospects, in *Agricultural Progress*, **54**, 1–17.

–, Law C and Miller T, 1991 *Wheat Yesterday, Today and Tomorrow*, Cambridge, Plant Breeding International & Plant Research Ltd.

Blackburn S, 1982 *The Life of a Norfolk Thatcher*, Salhouse, privately published.

Board of Agriculture & Fisheries, 1910 *Thatching*, Board of Agriculture & Fisheries Leaflet **236**, London, Board of Agriculture & Fisheries.

Boardman H C, 1933 Reed thatching in Norfolk, in *Architects Journal*, **77**, 563–7.

Bolton N and Chalkley B, 1990 The rural population turnaround: a case-study of north Devon, in *Journal of Rural Studies*, **6**, 29–43.

Bowick T, 1883 *The Crops of the Farm*, London, Bradbury Agnew.

Bowley M, 1960 *Innovations in Building Materials. An Economic Study*, London, Gerald Duckworth & Co.

Boys J, 1805 *General View of the Agriculture of the County of Kent; with Observations on the Means of its Improvement*, 2nd ed, London, Richard Phillips.

Bradshaw A, 1912 The thatching of ricks, in *Board of Agriculture Journal*, **19**, 301.

Brereton, C, 1991 *The Repair of Historic Buildings: advice on principles and methods*, London, English Heritage.

Brewer J G, 1972 *Enclosures and the Open Fields: A Bibliography*, London, British Agricultural History Society.

British Batavian Trading Company, 1915 *Thatching and How to Make it Permanently Fire-Proof*, London, The British Batavian Trading Company Limited.

Brock, D, 1999, The thatching years, *Conservation Bulletin*, **35**, 28–31.

Brockett, P, nd *Straw Thatching: When is Straw not a Straw?* unpublished report, private archive.

Brockett P and Wright A, 1986 *The Care and Repair of Thatched Roofs*, Technical Pamphlet **10**, London, Society for the Protection of Ancient Buildings & the Rural Development Commission

Brough S, 1976 A Cornish tradition: John Williams – Thatcher of Chacewater, in *Cornish Life*, **3**, 11–13.

Brown J, 1974 *A Thatcher's Memories*, typescript, Chelmsford Library.

–, 1978 Thatching with Father, in Seager 1978, 76–9.

Brown A G and Bradley C, 1995 Past and present alluvial wetlands and the eco-archaeo resource: implications from research in the East Midlands valleys (UK), in Rowley T (ed), The Evolution of Marshland Landscapes, Oxford, Rewley House, 283–95.

Burke J F, 1834–40 *British Husbandry*, 3 vols, London, Baldwin & Cradock.

Burn R S, 1878 [1889–1904] *Outlines of Modern Farming*, 5th edn, 5 vols, London, Crosby & Lockwood.

Burton A, 1891 *Rush-bearing: An Account of the Old Custom of Strewing Rushes; Carrying Rushes to Church; the Rush-cart; Garlands in Churches; Morris-Dancers; the Wakes; the Rush*, Manchester, Brook & Chrystal.

Caird J, 1851 *English Agriculture in 1850–51*, London, Longman & Co.

Campbell D, 1831 On thatching with fern, *Prize Essays and Transactions of the Highland Society of Scotland*, **2**, New Ser, 184–90.

Champion, A G, ed, 1989 *Counterurbanization: The Changing Pace and Nature of Popular Decentralization*, London, Edward Arnold.

Chapman V, 1982 Heather-thatched buildings in the northern Pennines, *Transactions of the Architectural & Archaeological Society of Durham & Northumberland*, **4**, new ser, 9–12.

Chapman W, 1798 Specification of the patent granted ... for a method of laying, twisting, or making ropes or cordages, in *Repertory of Arts and Manufactures*, **1**, 2nd ser, 1–44.

Charlton L, 1779 *The History of Whitby and of Whitby Abbey*, York.

Clapham A R, Tutin T G and Moore D M, 1987, *Flora of the British Isles*, 3rd edn, Cambridge, Cambridge University Press.

Clapp B W, Fischer H E S and Jurica A R J, 1977 *Documents in English Economic History*, 2 vols, London, G Bell.

Clark C, 1947 Thatch, thatchers and thatching, in *Agriculture*, **53**, 444–50.

Clark C, 1948 letter *The Times*, January.

Clifton-Taylor A 1972 *The Pattern of English Building*, London, B T Batsford.

Cohen, nd Roofs of reed, undated newspaper cutting, Ipswich Record Office

Collier J, 1831 On thatching with heath, *Prize Essays and Transactions of the Highland Society of Scotland*, **2**, new ser, 190–5.

Collins E J T, 1969 *Sickle to Combine. A Review of Harvest Techniques from 1800 to the Present Day*, Reading, Museum of English Rural Life.

–, 1970 *Harvest Technology and Labour Supply in Britain 1790–1870*, unpublished PhD thesis, University of Nottingham.

Collins W W, 1866 *Armadale*, 2 vols, London.

Coppock J T, 1964 *An Agricultural Atlas of England and Wales*, Faber, London.

–, 1971 *An Agricultural Geography of Great Britain*, London, Faber & Faber

Council for British Archaeology, 1985 *Making Sense of Buildings*, London, Council for British Archaeology.

Cowell J G, nd *Treatise on Thatching*, Soham, privately published.

Cowley W, 1955 The technique and terminology of stacking and thatching in Cleveland, in *Yorkshire Dialect Society Transactions*, **9**, pt 54, 35–40.

Cox J, 1994 *Thatch and Thatching from 1940*, 2 vols, internal report for English Heritage.

– and Thorp J, 1991 Authentic slating in Devon, in *Transactions of the Association for Studies in the Conservation of Historic Buildings*, **16**, 3–12.

Crampton A H, 1935 *Rushwork*, London, The Studio Ltd.

Crampton C and Mochrie E, 1931 *Rushwork*, 2nd edn, Leicester, Dryad Press.

Creasey J and Ward S, 1984 *The Countryside Between the Wars 1918–1940, A Photographic Record*, London, Batsford.

Crowther R E and Evans J, 1986 *Coppice*, 2nd edn, Forestry Commission Leaflet **83**, London, HMSO.

Darby H C (ed), 1976 *A New Historical Geography of England after 1600*, Cambridge, Cambridge University Press.

Darley G, 1978 *Villages of Vision*, London, Paladin.

Davey B J, 1980 *Ashwell 1830–1914. The Decline of the Village Community*, Occasional Paper **5**, 3rd Series, Leicester, Leicester University Press.

Davis T, 1811 *General View of the Agriculture of Wiltshire*, London, Board of Agriculture.

Deas J H, 1939 *Building in Norfolk*, unpublished thesis, Royal Institute of British Architects.

Department of the Environment, 1987 *Circular 8/87. Historic Buildings and Conservation Areas - Policy and Procedures* London, HMSO

Department of the Environment, 1994 *Planning Policy and Guidance: Planning and the Historic Environment*, **15**, London, HMSO

Dickes W F, 1906 *The Norwich School of Painting*, Norwich, Jarrold & Sons.

Dickson R W and Stevenson W A, 1815 *General View of the Agriculture of Lancashire with Observations on the Means of its Improvement*, London, Board of Agriculture.

Diplock A H, 1929 Sussex roofs, in *Sussex County Magazine*, **3**, 620–621.

Dodds L, 1929 Straw thatch as a building material, in *The National Builder*, October, 86–87.

Donaldson J, 1847 *The Cultivated Plants of the Farm*, London, R Groombridge and Sons.

Duckham, A N, Jewell A J, Fox S, Gibb J A C and Pearce J (eds), 1963 *Farming*, **4**, Caxton, London.

Duncan, R, 1947 *Home-Made Home*, London, Faber & Faber

Dundonald Earl of, 1794 Specification of the patent granted ... for his method of extracting or making tar, pitch, essential oils, volatile alkali, mineral acids, salts and cinders from pit-coal, in *Repertory of Arts and Manufactures*, **1**, 2nd ser, 145–8.

Dyer J, 1981 *My Early Days*, Twyford, privately published.

East Anglian Master Thatcher's Association, 1989 2nd ed *Technical Pamphlet*, **1**, East Anglian Master Thatcher's Association, Norwich.

Ellacott S E, 1981 *Braunton Farms and Farmers*, Aycliffe Press, Taunton.

Ellis C W, Eastwick-Field J and Eastwick-Field E, 1947 *Buildings in Cob, Pise and Stabilised Earth*, Country Life, London.

Elson M, 1959 Is the thatched roof doomed?, in *Country Life*, **126**, 526–8.

Emery N, 1985 The heather thatching of buildings, in *Making Sense of Buildings*, Council for British Archaeology, London, Council for British Archaeology, 89–94.

–, 1986 Fell Close Farm and the use of 'Black Thack', in *Durham Archaeological Journal*, **2**, 91–95.

English Heritage, 2000 *Thatch and Thatching*, English Heritage Guidance Note, London, English Heritage.

Environmental Appraisal Group, 1991 *Socio-Economic Impact of Changes in the Quality of Thatching Reed on the Future of the Reed-Growing and Thatching Industries and on the Wider Rural Economy*, School of Environmental Sciences, University of East Anglia

Evans E E, 1957 *Irish Folk Ways*, London, Routledge & Kegan Paul.

Farey J, 1811 *General View of the Agriculture and Minerals of Derbyshire with Observations on the Means of Their Improvement*, 3 vols, London, Board of Agriculture.

Farman, A, 1949 Retirement of reed thatcher, in *The Eastern Daily Press*, 3 January.

Fearn, J, 1972 *Thatch and Thatching*, Aylesbury, Shire.

Fenton A, 1978 *The Northern Isles, Orkney and Shetlands*, Edinburgh, Donald.

Fielden M E, 1934 Old-time survivals in Devon, *Report & Transactions of the Devonshire Association*, **66**, 357–373.

Filmer R, 1980 Kentish thatch, *Bygone Kent*, **1**, 324–331.

–, 1981 *Kentish Rural Crafts and Industries*, Rainham, Meresborough.

Fitzrandolph H E and Hay M D, 1926 *The Rural Industries of England & Wales. A Survey Made on behalf of the Agricultural Economics Research Institute Oxford. II. Osier-growing and Basketry and Some Rural Factories*, Oxford, Clarendon Press.

Foot P, 1794 *General View of the Agriculture of the County of Middlesex, with Observations on the Means of their Improvement*, London, Board of Agriculture.

Frankel O H, 1976 Natural variation and its conservation, in *Genetic Diversity in Plants Proceedings of an International Symposium on Genetic Control of Diversity in Plants held at Lahore, Pakistan, March 1–7, 1976*, Muhammed A, Aksel R and Borstal R C von, New York, Plenum Press.

Fyfe W W, 1863 Farm seeds and seeding, in *Journal of the Bath & West of England Society*, **11**, new ser, 296–353.

Gailey A, 1984 *Rural Houses of the North of Ireland*, Edinburgh, John Donald.

Gardiner D, 1949 *Companion into Dorset*, 4th edn, London, Methuen.

Gill N T and Vear K C, 1969 *Agricultural Botany*, 2nd edn, London, Duckworth.

Gooch W, 1811 *General View of the Agriculture of the County of Cambridge*, London, Richard Phillips.

Goodland N, 1953 *My Father Before Me*, London, Hutchinson.

Gratton H, 1936 letter, in *The Western Times*, October [cutting in West Country Studies Library].

Greenacre D W, 1958 Reeds for the thatcher, *The Field*, 24 April.

Grigg D, 1980 *Population Growth and Agrarian Change: An Historical Perspective*, Cambridge, Cambridge University Press.

–, 1989 *English Agriculture: An Historical Perspective*, Oxford, Basil Blackwell.

Grove, A T, 1962 Fenland, in *Great Britain: Geographical Essays*, Mitchell J B, Cambridge, Cambridge University Press, 104–122.

Gunn E, 1936 The art of the thatcher, *Somerset Countryman*, **6**:2, 30–31.

Halliwell W, 1905 English wheat and the development of British milling, *Journal of the Royal Agricultural Society*, **66**, 224–229.

Hansard 1948 *Oral Answers*, 15 April 1948.

Harman H, 1929 *Buckinghamshire Dialect*, London, Hazell, Watson & Viney.

Harper C G, 1921 Thatch, in *Journal of the American Institute of Architects*, **9**, 389–96.

Harris E and Savage N, 1990 *British Architectural Books and Writers 1556–1785*, Cambridge, Cambridge University Press.

Harrison J R, 1989 Some clay dabbins in Cumberland: their construction and form. Part I, in *Transactions of the Ancient Monuments Society*, **33**, 97–151.

Hartley, D, 1939 *Made in England*, London, Methuen.

Hartley M and Ingilby J, 1968 *Life and Tradition in the Yorkshire Dales*, London, Dent.

–, 1972 *Life in the Moorlands of North-East Yorkshire*, London, Dent.

–, 1978 Roofs gathered from nature. Ling thatches, in *Country Life*, **163**, 1022–4.

–, 1986 *Dales Memories*, London, The Dalesman.

Harvey J H, 1945 Mudtown, Walton-on-Thames, in *Surrey Archaeological Collections*, **49**, 127–9.

Haslam S M, 1972 *The Reed (Norfolk Reed)*, 2nd edn, Norwich, Norfolk Reed Growers Association.

–, Sinker C and Wolseley P, 1975 British water plants, in *Journal of Field Studies*, **4**, 243–351.

Hay A C de P, nd *St Michaels Church, Ingram: The Story*, Alnwick, privately published.

Hennell T B, 1934 *Change in the Farm*, Cambridge, Cambridge University Press.

Hervey-Murray C G, 1980 *The Identification of Cereal Varieties*, Cambridge, RHM Arable Services Ltd.

Hetrick, B A D and Wilson, G W, 1992 Mycorrhizal dependence of modern wheat vars, land races and ancestors, in *Canadian Journal of Botany*, **70**, 2032–2040.

Hickish, J R, 1960 Thatching costs a pretty penny, in *Farm and Country*, 3 August.

Hillman G C, 1983 *Archaeobotanical Criteria Used to Distinguish Tetraploid and Hexaploid Wheat Rachis Remains*, MSc course notes, London, Institute of Archaeology.

Hillyard C, 1837 *Practical Farming and Grazing*, 2nd edn, Northampton, T E Dicey.

Hine R, 1914 *The History of Beaminster*, Taunton, Barnicott & Pearce.

Hodgson J C (ed), 1914 Northern journeys of Bishop Richard Pococke, in *Publications of the Surtees Society*, **124**, 199–252.

Holden T G, 1998 *The Archaeology of Scottish Thatch*, Edinburgh, Historic Scotland Technical Advice Note **13**.

Holland H, 1808 *General View of the Agriculture of Cheshire*, London, Board of Agriculture.

Holt J, 1795 *General View of the Agriculture of the County of Lancaster: With Observations on the Means of its Improvement*, London, Board of Agriculture.

Hoppit, D, 1985 Thatching, *Traditional Homes*, January, 36.

Horrox D K, 1953 The folk who live at Ramsgill, *The Dalesman* **14**, 513–515.

Humphries A E, 1911 The milling of wheat in the United Kingdom, in *Journal of the Royal Agricultural Society*, **72**, 24–37.

Hunt T F, 1827 *Designs for Parsonage Houses, Alms Houses, etc*, London, Longman & Co.

Hutchins J, 1861-73 *The History and Antiquities of the County of Dorset*, 3rd edn, 4 vols, London, J B Nichols & Sons.

Innocent C F, 1916 *The Development of English Building Construction*, Cambridge, Cambridge University Press.

Isaac J W P, 1856 On the economical adaptation of existing agricultural dwellings to the health and comfort of the inhabitants, on improved sanitary principles, in *Journal of the Bath & West of England Society*, **4**, new ser, 111–125.

Jarman, R J and Pickett A A, 1994 *Botanical Descriptions of Cereal Varieties*, Cambridge, National Institute of Agricultural Botany.

Jekyll G, 1899 *Wood and Garden. Notes and Thoughts, Practical and Critical, of a Working Amateur*, London, Longman & Co.

Jenkins D T, nd *Indexes of the Fire Insurance Policies of the Sun Fire Office and the Royal Exchange Assurance 1775–1787*, typescript, Guildhall Library.

Jenkins J G, 1965 *Traditional Country Craftsmen*, London, Routledge & Kegan Paul.

Jobson A, 1949 The reed-thatcher's tools, *Country Life*, **105**, 1187.

–, 1961 The thatched churches of Suffolk, in *East Anglian Magazine*, **20**, 494–504.

Johnson S, 1827 *Dictionary of the English Language*, London.

Johnston J, 1847 *Lectures on Agricultural Chemistry and Geology*, Edinburgh, William Blackwood & Sons.

Jones A M, 1927 *The Rural Industries of England & Wales. A Survey Made on behalf of the Agricultural Economics Research Institute Oxford. IV. Wales*, Oxford, Oxford University Press.

Jones E L, 1968 The reduction of fire damage in southern England, 1650–1850, in *Post-Medieval Archaeology*, **2**, 140–149.

– and Falkus M E, 1990 Urban improvement and the English economy in the seventeenth and eighteenth centuries, in Borsay P (ed), 1990 *The Eighteenth-Century Town; A Reader in English Urban History 1688-1820*, London, Longman, 119–127.

Juniper, B E, 1990 Straw: its structure, chemistry and the possibilities for its further use as a raw material for industry, in *Agricultural Progress*, **65**, 23–38

Katz H R, 1976 *The Decline of Thatch*, unpublished thesis, Architectural Association.

Kennaway L M, nd *To Lovers of English Rural Scenery*, privately published.

Kent N, 1796 *General View of the Agriculture of the County of Norfolk; With Observations For the Means of its Improvement*, London, Board of Agriculture.

Kirby J J, Marigold E A and Ansell M P, 1990 *The Quality of Combed Wheat Reed for Thatching*, unpublished manuscript, private collection.

Kirby J J and Rayner A D, 1986 *The Biodegradation of Thatching Straw*, typescript, project reference CSA 900, School of Biological Sciences, University of Bath.

–, 1988 Disturbance, decomposition and patchiness in thatch, in *Proceedings of the Royal Society Edinburgh*, **94B**, 145–153.

–, 1989 *Aspects of the Decomposition, Mechanical Strength and Anatomy of Water Reed (P. Australis) used in Thatching*, MAFF Report, School of Biological Science, University of Bath.

Lambert J M, Jennings J N, Smith C T, Green C and Hutchinson J N, 1960 *The Making of the Broads. A Reconsideration of Their Origin in the Light of New Evidence*, London, John Murray.

Langton J and Morris R J, 1986 *An Atlas of Industrializing Britain, 1780–1914*, London, Methuen.

Lankester E, 1832 *The Library of Entertaining Knowledge. Vegetable substances used for the food of man*, London, Charles Knight.

Laugier M A, 1755 *An Essay on Architecture*, London, T Osborne & Shipton.

Lawton R and Podley C G, 1992 *Britain 1740–1950: An Historical Geography*, London, Edward Arnold.

Laycock C H, 1920 The old Devon farm-house. Part I. Its exterior aspect and general construction, in *Report & Transactions of the Devonshire Association*, **52**, 158–91.

Le Couteur J, 1836 *On the Varieties, Properties and Classification of Wheat*, Jersey, privately published.

Le Vegetal, 1977 *Les toits dans le paysage*, Strasbourg, Maian Mane Clarne, Imprimirie Istra.

Letts J, 1999 *Smoke-Blackened Thatch (SBT). A Unique Source of Late Medieval Plant Remains from Southern England, London & Reading*, London, English Heritage & the University of Reading.

Lisle E, 1757 *Observations in Husbandry*, Faulkner, Dublin.

Loudon J C, 1831 *An Encyclopedia of Agriculture*, 2nd edn, London, Longman, Rees, Orme, Brown & Green.

Low D, 1843 *Elements of Practical Agriculture*, London, Longman, Brown, Green & Longmans.

Lowe J, 1994 *The Dorset Thatching Report*, typescript, Dorset County Council.

Lucas A T, 1960 *Furze. A Survey and History of its Uses in Ireland*, Dublin, Educational Company of Ireland.

Lucas R, 1995 Some observations on descriptions of parsonage buildings made in Norfolk glebe terriers, in *Transactions of the Ancient Monuments Society*, **39**, 85–98.

Lupton F G H, 1987 The history of wheat breeding, in Lupton F G H, *Wheat Breeding: its Scientific Basis*, London, Chapman & Hall, 52–71.

Lyall S, 1988 *Dream Cottages. From Cottage Ornée to Stockbroker Tudor. Two Hundred Years of the Cult of the Vernacular*, London, Robert Hale.

Machin R, 1994 *Rural Housing: An Historical Approach*, London, The Historical Association.

McDonnell J (ed), 1963 *A History of Helmsley, Rievaulx and District*, York, Stonegate Press.

McDougall D S, 1958 Harvesting the Norfolk reed, in *The Field*, 21 March, 493–4.

McGlue, nd Separate staircases and thatch over all, *Traditional Homes*, 5.

Mackenzie E, 1825 *An Historical, Topographical, and Descriptive View of the County of Northumberland*, 2nd edn, 2 vols, Newcastle upon Tyne.

McLaren, G, 1991 *The Conservation of Thatched Buildings in Great Britain*, unpublished diploma thesis, the Architectural Association.

Malcolm W, Malcolm James and Malcolm Jacob, 1794 *General View of the Agriculture of the County of Surrey, with Observations on the Means of its Improvement*, London, Board of Agriculture.

Manners J E, 1979 Roofs for a poor man. Thatching in Ireland, in *Country Life*, **166**, 271–2.

Marshall W, 1790 *The Rural Economy of the West of England: including Devonshire; and Parts of Somersetshire, Dorsetshire, and Cornwall*, 2 vols, London, Nicol.

–, 1818 The Midlands, *Review and Abstracts of the County Records of the Board of Agriculture*, **4**, London, David & Charles.

Martin D and Martin B, 1978 *Historic Buildings in Eastern Sussex Rape of Hastings Architectural Survey*, **2**.

–, 1979 *Historic Buildings in Eastern Sussex Rape of Hastings Architectural Survey*, **4**.

Massingham H J, 1943 *Men of Earth*, London, Chapman & Hall.

Mathias P, 1969 *The First Industrial Nation: An Economic History of Britain, 1700–1914*, London, Methuen.

Maufe, E, 1946 *The Architectural use of Building Materials*. Post-War Building Studies **18**, London, HMSO.

Mavor W, 1809 *General View of the Agriculture of Berkshire*, London, Richard Phillips.

Meeson R A and Welch C M, 1993 Earthfast posts: The persistence of alternative building techniques, in *Vernacular Architecture*, **24**, 1–17.

Meirion-Jones G I, 1976 Some early and primitive building forms in Brittany, in *Folk Life*, **14**, 46–64.

Metcalfe A, 1953 It's warm under the thatch, in *The Dalesman*, **15**, 296–7.

Middleton J, 1798 *View of the Agriculture of Middlesex; With Observations on the Means of its Improvement and Several Essays on Agriculture in General*, London, Board of Agriculture.

–, 1800 *Experiments and Observations on Various Kinds of Manure*, Lambeth, privately published.

Ministry of Agriculture ('John Fallowfield'), 1941 *Harvesting Problems and the Modern Combine: Does the Modern Combine Save Money? An Official Enquirers Report*, London, HMSO.

Mitchell B R and Deane P, 1962 *Abstract of British Historical Statistics*, Cambridge, Cambridge University Press.

Mitchell T J, 1961a Yorkshire thatch, in *The Dalesman*, **22**, 783–788.

–, 1961b Yorkshire thatch, in *The Dalesman*, **23**, 134–135.

Moir J, 1990 *A World unto Themselves? Squatter Settlement in Herefordshire 1780–1880*, unpublished PhD thesis, University of Leicester.

Moir J and Letts J B, 1999 *Thatch: Thatching in England 1790–1940*, English Heritage Research Transactions Series 5, London, English Heritage.

Moore A W, 1982 *John Sell Cotman, 1782–1842*, Norwich, Norfolk Museum Service.

Morgan W E and Cooper F W, 1960 *The Thatcher's Craft*, Salisbury, Rural Industries Bureau.

Morris G L, 1909 The home – III. A thatched cottage, in *The British Architect*, **71**, 149–150.

Morrison J (ed), 1898 *The Diaries of Jeffrey Whitaker 1739–1741*, Wiltshire Record Society **44**, Salisbury.

Morrison J, 1993 *Corn Varieties Grown in the Nineteenth Century At Grants Farm, Bratton, Wiltshire*, typescript, private collection.

Morton J, 1862 *The Farmer's Calendar*, 21st edn, London, Routledge, Warne & Routledge.

Murray J, 1895 *Handbook for Hertfordshire, Bedfordshire and Huntingdonshire*, London, John Murray.

Nash, J, 1991, *Thatchers and Thatching* London, B T Batsford

Nash J, 1994 Harold Wright, in *Somerset Magazine*, **4**, 16–19.

Neve R, 1726 [1969] *The City and Country Purchaser and Builder's Dictionary*, London, David and Charles.

Nightingale F W, 1939 Must a beautiful rural craft die?, in *Northampton Independent*, 24 March, Supplement, iv.

Northamptonshire County Council, 1935 Rural housing, in *Annual Report*, Northampton, Northamptonshire County Council.

Nugat P, 1950 James Fosberry – thatcher, in *The Sussex County Magazine*, **24**, 245–247.

Oldershaw A W, 1944 *Good Farm Crops*, London, Hodder & Stoughton.

– and Porter J, 1929 *British Farm Crops*, London, Ernest Benn.

Oxford English Dictionary, 1989, Oxford, Oxford University Press.

Palmer J, 1978 Craftsmen then and now: the thatcher, in *Northamptonshire Life*, January 1978, 40–43.

Palmer J D, 1970 Plant breeding today, in *Journal of the Royal Agricultural Society*, **131**, 7–17.

Parkinson R, 1808 *General View of the Agriculture of the County of Rutland, with Observations on the Means of its Improvement*, London, Board of Agriculture.

–, 1811 *General View of the Agriculture of the County of Huntingdon*, London, Richard Phillips.

Parrott W R, 1974 *Sixty Years a Thatcher*, Clapham, Beds, privately published.

Partridge M, 1973 *Farm Tools through the Ages*, Reading, Osprey Publishing.

Patterson W G R (ed), 1925 *Farm Crops*, 4 vols, London, Gresham Publishing.

Peachey R A, 1951 *Cereal Varieties in Great Britain*, London, Crosby Lockwood & Son.

Peate I C, 1944 *The Welsh House. A Study in Folk Culture*, 2nd edn, Liverpool, H Evans & Sons.

Percival J, 1921 *The Wheat Plant: A Monograph*, London, Duckworth & Co.

–, 1942, *Agricultural Botany*, 8th edn, London, Duckworth.

–, 1943 *Wheat in Great Britain*, Shinfield, London, Duckworth.

Peters J E C, 1977 The solid thatch roof, in *Vernacular Architecture*, **8**, 825.

Phillips A D M, 1989 *The Underdraining of Farmland in England During the Nineteenth Century*, Cambridge, Cambridge University Press.

Phillips N J A, 1975 *Dykes of Romney Marsh*, Hawkhurst, privately published.

Pitt W, 1809a *General View of the Agriculture of the County of Northampton*, London, Board of Agriculture.

–, 1809b *A General View of the Agriculture of the County of Leicester; with Observations on the Means of its Improvement; to which is annexed a survey of the county of Rutland*, London, Board of Agriculture.

–, 1810 (1969) *General View of the Agriculture of the County of Worcester*, David & Charles

–, 1813 *General View of the Agriculture of the County of Stafford; with Observations on the Means of its Improvement*, London, Board of Agriculture.

Plaw J, 1800 *Sketches for Country Houses, Villas, and Rural Dwellings*, London, J Taylor.

Plymley J, 1803 *General View of the Agriculture of Shropshire: With Observations*, London, Board of Agriculture.

Pocock W F, 1807 *Architectural Designs for Rustic Cottages, Picturesque Dwellings, Villas*, London, J Taylor.

Porter S, 1986 Thatching in early-modern Norwich, in *Norfolk Archaeology*, **39**, 310–312.

Potter T, 1914 Roof coverings – thatch, in *Journal of the Society of Estate Clerks of Works*, **27**, 41–49.

Powell A H, 1923 Modern craftsmanship. 5 – Thatching, in *The Architects Journal*, **58**, 859–862.

Price U, 1794 [1871] *Essays on the Picturesque*, Farnborough.

Priest St J, 1810 *General View of the Agriculture of Buckinghamshire*, London, Richard Phillips.

Prothero R E, 1901 English agriculture in the reign of Queen Victoria, in *Journal of the Royal Agricultural Society*, **62**, 1–39.

–, 1923 *The Land and its People; Chapters in Rural Life and History*, London, Hutchinson & Co.

Prufrock, 1971 Since when were thatchers a Whitehall pressure group?, in *The Sunday Times*, 14 November.

Pullen, J H, 1979 *The Production of Wheat Reed for Thatching*, unpublished ms, MAFF (ADAS) Cullompton, Devon.

Purseglove J, 1988 *Taming the Flood, A History and Natural History of Rivers and Wetlands*, Oxford, Oxford University Press with Channel Four.

Rackham O, 1980 *Ancient Woodland, its History, Vegetation and Uses in England*, London, Arnold.
–, 1986a *The History of the Countryside*, London, Dent.
–, 1986b *Marshes, Fens, Rivers and Sea* London, Dent
Ransome J A, 1843 *The Implements of Agriculture*, London, J Ridgway.
Raynes H E, 1964 *A History of British Insurance*, 2nd edn, London, Isaac Pitman & Sons.
Rea J T, 1941 *How to Estimate, being the Analysis of Builders' Prices*, 6th edn, London, Batsford.
Rham W L, 1845 *The Dictionary of the Farm*, 2nd edn, London, Charles Knight & Co.
Ricauti T, 1848 *Sketches for Rustic Work*, London, Henry G Bohn.
Ridsdale F J, 1946 Thatched houses, in *The Dalesman*, **8**, 138.
Robertson A J, 1938 *A History of Alresford, Derived from Manuscript Notes by Robert Boys*, Winchester, Warren & Son.
Robinson, M, & Lambrick, G, 1989 Holocene alluviation and hydrology in the Upper Thames Basin, in *Nature* **308**, 809–814.
Roffey M and Cross C, 1933 *Rush-Work*, London, Pitman.
Rogers J, 1976 *In the Life of a Country Thatcher*, Modbury, Modbury Local History Society.
Rowley T (ed), 1981 *The Evolution of Marshland Landscapes, Papers Presented to A Conference Held in Oxford in December 1979*, Oxford, Oxford University Department for Extramural Studies.
Royal Commission on the Historical Monuments of England, 1987 *Houses of the North York Moors*, London, HMSO.
Royal Society for the Protection of Birds, 1994 *Reedbed management for bitterns* Sandy, RSPB.
Rudge T, 1813 *General View of the Agriculture of the County of Gloucester*, London, Richard Phillips.
Rural Industries Bureau, 1939–47, 1947–1968, 1951–62 *Annual Reports*, Salisbury, Rural Industries Bureau.
Rural Industries Bureau, 1965 *Thatching. A Report on a Survey conducted to Assess the Value of Thatching as a National Asset, on the Present State of the Craft and Factors Affecting it*, Salisbury, Rural Industries Bureau.
Rural Industries Bureau, 1971

Salmon S C, 1964 *The Principles and Practice of Agricultural Research*, London, Leonard Hill.
Salzman, L F, 1952 *Building in England Down to 1540. A Documentary History*, London, Oxford University Press
Sandon E, 1977 *Suffolk Houses. A Study of Domestic Architecture*, Woodbridge, Baron Publishing.
Report of the Committee on Land Utilisation in Rural Areas, 1942 London, HMSO.
Seager, E (ed), 1978 *The Countryman Book of Village Trades and Crafts*, Burford, The Countryman.
Seddon, Q, 1989 *The Silent Revolution*, London, BBC Books.
Seymour J, 1956 St Day man carrying on almost a lost art, *West Briton*, 4 December.
Seymour J, 1984 *The Forgotten Arts*, London, Dorling Kindersley.
Smedley N, 1976 *Life and Tradition in Suffolk & North-East Essex*, London, Dent.
Snell K D M, 1985 *Annals of the Labouring Poor: Social Change and Agrarian England, 1660-1900*, Cambridge, Cambridge University Press.
Souness J R, 1992 Taighean tugha tirisdeach: The thatched houses of Tiree, in Riches A and Stell G (eds), 1992 *Materials and Traditions in Scottish Building, Essays in Memory of Sonia Hackett*, Regional & Thematic Studies **2** Edinburgh, Scottish Vernacular Buildings Working Group, 81–96.
Stanford, C, 1994 *Results of Questionnaire on Thatching*, unpublished manuscript, private collection.
Staniforth, A R, 1979 *Cereal Straw*, Oxford, Oxford University Press.
Stedman A R, 1960 *Marlborough and the Upper Kennet Country*, Marlborough, privately published.
Stephens W B, 1908 *Book of the Farm*, 5th edn, 3 vols, Edinburgh, William Blackwood & Sons.
Stevenson W, 1809 *General View of the Agriculture of the County of Surrey*, London, Board of Agriculture.
–, 1812 *General View of the Agriculture of the County of Dorset: with Observations on the Means of its Improvement*, London, Board of Agriculture.
Storer B, 1985 *The Natural History of the Somerset Levels* Wimborne, The Dovecote Press
Stranks C, 1990 *Warmington Remembered: A Warwickshire Village and its People, 1915–1990*, Warmington, Fir Tree.
Street A G, 1933 Thatching and thatchers, in *The Listener*, **9**, 753–4.
Strickland H E, 1812 *A General View of the Agriculture of the East Riding of Yorkshire*, York, Board of Agriculture.
Sutton A (ed), 1991 *Cotswold Tales*, Stroud, Sutton.
Sykes J D, 1981 Agriculture and science, in Mingay G E (ed), 1981 *The Victorian Countryside*, **1**, London, Routledge & Kegan Paul, 260–72.

Thacker F S, 1932 *Kennet Country*, Oxford, Basil Blackwell.
Thomas W E, 1940 Thatching in Dorset, in *Journal of the Ministry of Agriculture*, **46**, 468–472.
Tuke J, 1800 *General View of the Agriculture of the North Riding of Yorkshire*, London, Board of Agriculture.

University of East Anglia, 1991 *Socio-Economic Impact of Changes in the Quality of Thatching Reed on the Future of the Reed-Growing and Thatching Industries and on the Wider Rural Economy*, University of East Anglia, Norwich.

Vancouver C, 1808 *General View of the Agriculture of the County of Devon; with Observations on the means of its Improvement*, London, Board of Agriculture.
–, 1813 *General View of the Agriculture of Hampshire, including the Isle of Wight*, London, Board of Agriculture.
Vilmorin-Andrieux et Cie, [1880] *Les meilleurs blés. Description et culture des principales variétés de froments d'hiver et de printemps*, Paris, Vilmorin-Andrieux et Cie.
Vincent J, 1941 Harvesting the reeds, in *Country Life*, **90**, 1119.

Walker B, McGregor C and Stark G, 1996 *Thatches and Thatching Techniques. A Guide to Conserving Scottish Thatching Traditions*, Edinburgh, Historic Scotland.
Walton J, 1975 The English stone-slaters' craft, in *Folk Life*, **13**, 38–53.
Ward & Lock, 1880 *Ward & Lock's Book of Farm Management and Country Life*, London, Ward, Lock & Co.
Ward J D U, 1939 Some thatched roofs, in *Country Life*, **86**, 431.
–, 1946 The future of thatch, in *Country Life*, 20 September, 529.
–, 1960 Thatched churches of East Anglia, in *Country Life*, **128**, 1508–11.
–, 1963 Thatched churches, in *Town and Country Planning*, **31**, 226–9.
Watkins M, 1981 *The English, the Countryside and its People*, London, Elm Tree.
Watson R C and McClintock M E, 1979 *Traditional Houses of the Fylde*, Centre for North-West Regional Studies, University of Lancaster Occasional Paper **6**, Lancaster, University of Lancaster.

Wedge T, 1794 *General View of the Agriculture of the County Palatine of Chester. With Observations on the Means of its Improvement*, London, Board of Agriculture.

West R C, 1987 *Thatch: A Manual for Owners, Surveyors, Architects and Builders*, Newton Abbot, David & Charles.

Whitlock J, 1960 Thatch as thatch can, in *The Field*, 2 June, 1073.

Williams W M, 1958, *The Country Craftsman. A Study of Some Rural Crafts and the Rural Industries*, London, Routledge & Kegan Paul.

Wilson J, 1862 *British Farming. A Description of The Mixed Husbandry of Great Britain*, Adam & Charles Black, Edinburgh.

Wilson J M, 1871 *The Farmer's Dictionary; or, A Cyclopedia of Agriculture*, 2 vols, Fullarton, Edinburgh.

Wood M P, 1921 Sussex thatch, in *The Builder*, **121**, 544.

Woodforde J 1969 (1979) *The Truth About Cottages*, Routledge Kegan Paul, London.

Wood-Jones R B, 1963 *Traditional Domestic Architecture of the Banbury Region*, Manchester, Manchester University Press.

Woods S H, 1988 *Dartmoor Stone*, Exeter, Devon Books.

Woodward D (ed) 1984 *The Farming and Memorandum Books of Henry Best, of Elmswell, 1642* London, Oxford University Press.

Worgan G B, 1811 *General View of the Agriculture of the County of Cornwall*, London, Board of Agriculture.

Worlidge J, 1694 *Mr Worlidges Two Treatises: the first, of improvement of husbandry ... the second, a treatise of cyder*, London.

Wright J (ed), 1905 *The English Dialect Dictionary*, 8 vols, Oxford, Henry Frowde.

Wright R P (ed), 1891 *Principles of Agriculture*, London, W G Blackie & Son.

Wrigley E A, 1988 *Continuity, Chance and Change. The Character of the Industrial Revolution in England*, Cambridge, Cambridge University Press.

Wymer, N, 1946 *English Country Crafts*, London, B T Batsford

Young A, 1804a *General View of the Agriculture of Hertfordshire*, London, Macmillan.

–, 1804b *General View of the Agriculture of the County of Norfolk*, London, Macmillan.

–, 1807 *General View of the Agriculture of the County of Essex*, 2 vols, London, Macmillan.

–, 1813a *General View of the Agriculture of the County of Suffolk*, London, Macmillan.

–, 1813b *General View of the Agriculture of the County of Lincolnshire*, 2nd edn, London, Macmillan.

Young, A W, & Davies, D, 1990 Anatomical investigations of Common Reed Phragmites Australis (CAV) Trin. ex Steudel and Flote grass Glyceria fluitans (L.) R. Br. found in different habitats in the British Isles, in *Agricultural Progress* **65**, 12–22.

Zohary D and Hopf M, 1988 *Domestication of Plants in the Old World, the Origin and Spread of Cultivated Plants in West Asia, Europe and the Nile Valley*, Oxford, Clarendon Press.

MANUSCRIPT SOURCES

Public Record Office

1: D4/421 Rural Industries Inquiry, 1930/1
2: MAFF 33/768
3: MAFF 33/770 6510 Executive Committee Meetings of the Rural Industries Bureau 1941-44
4: MAFF 33/770 6510 'D Straw Thatchers'
5: MAFF 33/772 6510 Thatching Instruction, Rural Industries Bureau
6: MAFF 33/772 6510 Training Arrangements
7: MAFF 113/523 Society for the Protection of Ancient Buildings Conference on housing, 1958
8: MAFF 113/522 Rural Industries Bureau Council Minutes and Papers, 1957-8
9: MAFF 113/99 XC6552
10: D1231 7 December 1966
11: MAFF 113/522 Draft of 1957-8 Rural Industries Bureau Annual Report
12: MAFF 113/98 XC6552 Minutes of Thatchers' Conference
13: D4/903 1953 survey of reed cutting
14: MAFF 113/103 Draft of 1954-5 Rural Industries Bureau Annual Report
15: D4/1200
16: MAFF 113/99 XC6552 Letter from Ruth Pollock, 26 October 1950
17: MAFF 113/523 Analysis of returns received by January 1960, to the Rural Industries Bureau's questionnaire on thatched properties
18: D4/12 7 3

Personal communications

Brockett, P; thatcher, instructor (Bedfordshire)
Cater, C; reed producer (Norfolk)
Cleeve, S; thatcher (Hampshire)
Cousins, J; thatcher (Suffolk)
Davis, K; thatcher (Oxfordshire)
Death, J and B; thatchers (Essex)
Dodson, M and A; thatchers (Huntingdonshire)
Dodson, J; thatcher (Huntingdonshire & Devon)
Dray, M; thatcher (Devon)
Dunkley, G; thatcher (Northamptonshire)
Elston, G; thatcher (Devon & Cornwall)
Evans, P; The Rural Development Commission
Farman, D; thatcher (Norfolk)
Fuchs, A; thatcher (Dorset)
Ganly, M; farmer (Oxfordshire)
Gendall, R; Cornish linguist, (Devon)
Glover, G; historian of agricultural machinery (Devon)
Godfrey, M; straw grower (Berkshire)
Goodacre, J; historian (Leicestershire)
Greenhill, G; curator (Lincolnshire)
Hall, A; archaeobotanist (Yorkshire)
Handley, T & A; rush chair makers (Oxfordshire)
Hannabuss, A; thatcher (Devon)
Haslam, S; botanist (Cambridgeshire)
Howard, A, cob builder (Devon)
Johnson, T; thatcher (Devon)
Kerrou, M; conservation officer (Northamptonshire)
King, J; thatcher, instructor
Lewis, A; thatcher (Wiltshire)
Lowe, J; conservation officer (Dorset)
Miller, R, thatcher, reed dealer (Dorset)
Mustiere, N; thatcher (Brittany, France)
Norman, P; thatcher, instructor (Somerset)
Pearce, S; thatcher (Wiltshire)
Pitt-Rivers; estate owner (Dorset)
Prince, A; thatcher (Devon)
Pydyn, A; archaeologist (Oxfordshire)
Slocombe, P; building historian (Wiltshire)
Snowdon, H; farmer (Devon)
Thomas, B; art historian (Oxfordshire)
Trezise, D; thatcher (Devon)
White, C; thatcher (Buckinghamshire)
White, I; straw grower (Devon)
Whiteley, T; thatcher (Dorset)
Wisbey, D; thatcher, instructor
Wright, H; straw grower, thatcher (Somerset)
Wright, R; straw grower, thatcher (Somerset)